The Uncanny Valley

IN GAMES & ANIMATION

The Uncanny Valley
IN GAMES & ANIMATION

Angela TINWELL

CRC Press
Taylor & Francis Group
Boca Raton London New York

CRC Press is an imprint of the
Taylor & Francis Group, an **informa** business

AN A K PETERS BOOK

CRC Press
Taylor & Francis Group
6000 Broken Sound Parkway NW, Suite 300
Boca Raton, FL 33487-2742

Printed on acid-free paper
Version Date: 20141020

International Standard Book Number-13: 978-1-4665-8694-9 (Hardback)

Library of Congress Cataloging-in-Publication Data

Tinwell, Angela.
   The uncanny valley in games and animation / Angela Tinwell.
      pages cm
      Summary: "This book is based on a series of empirical studies the author conducted to establish how aspects of facial expression and speech may be manipulated to control the Uncanny Valley in character design. It presents a novel theory that goes beyond previous research in that the cause of the Uncanny Valley is based on a perceived lack of empathy in a character. This book makes an original, scholarly contribution to our current understanding of the Uncanny Valley phenomenon and fills a gap in the literature by assessing the biological and social roots of the uncanny and its implications for computer-graphics animation. "-- Provided by publisher.
      Includes bibliographical references and index.
      ISBN 978-1-4665-8694-9 (hardback)
      1. Computer animation. 2. Computer graphics. 3. Computer games--Design. 4. Computer games--Social aspects. 5. Human engineering. 6. Human mechanics--Computer simulation. 7. Human body--Computer simulation. I. Title.

   TR897.7.T597 2014
   006.6'96--dc23   2014024617

Visit the Taylor & Francis Web site at
http://www.taylorandfrancis.com

and the CRC Press Web site at
http://www.crcpress.com

Contents

Acknowledgments

I'd like to express my gratitude to my family and friends who have offered continued support and patience, not only in the authorship of this book but also the years prior when I was conducting my research on the Uncanny Valley.

To my mother, Josephine Tinwell, who has remained an anchor on this journey, and my father, Peter Tinwell, who has been the compass that I have followed: thank you both for your help and wisdom.

My thanks go to editor Rick Adams, who had faith in me to achieve the book, even when I doubted my ability for completion.

I am grateful to all the participants who took part in my studies, without which this research and book would not be possible. Thank you for your time and contribution.

My appreciation is given to Roy Attwood, who helped with organizing the fantastic actors and video equipment for stimuli used in my experiments.

My warmest thanks go to the gifted animator Dr. Robin J. S. Sloan for his generosity and help in granting me permission to use his animations that signify a near lifetime of research into facial expression and characters. Robin's magnanimous and noble approach has made this book and other research projects a worthwhile and pleasurable project to be involved in.

Lance Wilkinson, the talented 3D artist behind many of the images in this book, has provided a delight for all with his excellent artwork, and I thank him for sharing.

Thanks go to my twin sister, Dr. Claire Tinwell, who, despite a busy schedule of her own, is always happy to discuss my research and offer her advice and new ideas.

My gratitude goes to Professor Tara Brabazon, for her guidance about the publishing process and the proposal for this book. I am so appreciative for her time and expertise.

My thanks go to the truly inspirational and motivational teacher Rebecca Ryder, whose excellence she so willingly shares with others.

Thanks to Professor Frank Pollick for his advice, debate, and insights about the uncanny.

Associate Professor Karl F. MacDorman's thought-provoking works on the Uncanny Valley in robots instilled motivation for my research, so I am grateful to him for his exceptional work, a source of inspiration for many.

Thanks also to Dr. Deborah Abdel Nabi for her perspective and opinion on the uncanny and advice on psychological concepts.

Masahiro Mori's beautiful mind, with a simple diagram of the Uncanny Valley, triggered profound research and debate for others to follow.

Finally, my thanks go to Professor Mark Grimshaw, my doctorate supervisor with whom I started this research journey. I will be forever grateful for his insightful feedback and guidance.

Author Biography

Dr. Angela Tinwell's research on the Uncanny Valley in human-like characters is recognized at an international level. As well as British media coverage on BBC television and radio, her work has been featured in news articles for *The Guardian* and *Times Higher Education* and in the American magazines *Smithsonian*, *New Yorker*, and *IEEE Spectrum Magazine*. In 2012 Tinwell completed her PhD dissertation, titled "Viewer Perception of Facial Expression and Speech and the Uncanny Valley in Human-Like Virtual Characters," and she has since published extensive studies on the topic. Her publications include empirical studies in the journal *Computers in Human Behavior* and theoretical writings for Oxford University Press. Tinwell's research into the Uncanny Valley in human-like characters is relevant in academia and industry, and she has presented her work with animators from the special effects company Framestore at the London Science Museum. As part of the Digital Human League, Tinwell is working with visual effects professionals at Chaos Group (creators of V-Ray rendering software) aimed at overcoming the Uncanny Valley.

Introduction

Human-Like Characters in Games and Animation

A s a senior lecturer in the areas of games and animation, I was fascinated by the developments made and controversy raised in creating realistic, human-like characters for games and animation. Particular criticism was made of the inability of these characters to clearly communicate facial expression of emotion to the viewer, possibly due to the technological challenges of achieving this. Their reported awkward and strange facial expression did not match the emotive qualities of the character's speech or the context of the scene in which the character was placed (Crigger, 2010; Doerr, 2007). Given that the majority of us rely on the effective communication of facial expression in ourselves and others as part of social interaction on a daily basis, theoretically one may presume that attempting to model such expressions and behavior in a virtual character would be a straightforward, intuitive process. Yet observation of my students struggling to model facial expression in characters with a realistic human-like appearance and, importantly, to effectively animate such expressions so that the character was regarded as authentically and believably human challenged this presumption. Working with my students to attempt to make their character designs more realistic and believable, I was reminded of the multidimensional qualities of facial expression and the complexities of human emotion. In addition to a specific facial expression of emotion (such as anger or happiness), the physiology of the face allows for different emotions to be portrayed independently in different areas of the face at the same time. Consideration should also be given to the onset and offset timings of expressions that may relate to the intensity of the emotion experienced, thus demonstrating the complexities of the emotion process. Designers may get away with (albeit benefit from) an oversimplification of facial expression in anthropomorphic-type characters with a cartoonish, stylized appearance, yet this approach could clearly not be taken with characters of a realistic, human-like appearance.

Much of the advice that I gave to students was to improve how one facial expression may develop and then transform to the next expression, a process that I referred to as *expression choreography*. However, how does one define this concept? Furthermore, what would be the implications of a character's speech on their facial expression and how that expression shifted to the next? This difficulty not only in creating but also in *explaining* how to represent emotional expression in human-like virtual characters suggested to me that our current understanding of how to define and represent facial expression of emotion needed to be improved. At first, I looked to examples of near-realistic human-like characters featured in contemporary games and animation to see how this dilemma may be resolved. Yet it appeared that even professionals working in the animation and games industry may not have found the answer to this plight.

Advances in technology have enabled animators and video game designers to pursue a high degree of simulated realism and include realistic, human-like characters in animation and games (e.g., in genres such as action-adventure games and role-playing games). It was intended that this increased realism would allow the viewer to appreciate the emotional state of the characters, thus leading to a heightened state of engagement (and enjoyment) in the game or animation (Doerr, 2007; Hoggins, 2010; Ravaja et al., 2008). However, rather than accepting these new realistic, human-like characters, the audience was critical of them (Crigger, 2010; Doerr, 2007; Hoggins, 2010). Characters that fell below viewer expectations in terms of their appearance and behavior were associated with the Uncanny Valley (Mori, 1970/2012). The Uncanny Valley, which proposes that viewers are generally less accepting of a character as the human-likeness for that character increases, now exists as a hypothetical benchmark for measuring whether a character is perceived as believably realistic and authentically human-like. The words *realism* and *realistic* are used, for the purposes of this study, to indicate that a designer has intended that a character will be perceived as authentically and believably human-like (Sommerseth, 2007). The perceived realism and human-likeness of a character may be judged on aspects such as a character's appearance, movement, behavior, sound and context and the relationship between all of these factors.

To create a character's facial expression and body movements in pre-recorded animation and games footage such as full-motion video (FMV) game trailers or cut scenes, facial and body motion capture techniques are combined with postproduction editing in 3D software. The process of motion capture involves multiple digital cameras that capture an actor's facial expression and body movements from precision-based markers placed on their body, such as their facial features or knee joint. These data are then transferred to a 3D character model on a computer and are used to animate that character. A designer may then further edit this footage frame by frame using 3D animation software to help ensure that the scene is as realistic as possible. However, it may be argued that video game designers face greater technological challenges in attempting to overcome the Uncanny Valley as in-game footage is played in real time. Instead they must rely on automated and/or procedural generation techniques in game engine software to simulate realism in game. As such, a character's facial expression, speech and body movement that is rendered in real time may lack the detail of that achieved in prerecorded animation. This factor was emphasized when, in 2006, much attention in the gaming community was focused on the Electronic Entertainment Expo (E306) where Quantic Dream was scheduled to show a technical demonstration of footage rendered in real time using the PS3 graphics engine. This short film, entitled "The Casting," allowed the audience a glimpse of what would await them in the much-anticipated new title *Heavy Rain*. Promoted as a cinematic crime-thriller game, Quantic Dream boasted that, using its new game engine technologies for simulating realism, it had developed a believably authentic human-like character capable of evoking contingent emotion in others. In other words, not only would the viewer be able to understand and relate to this character's emotion, but he/she, too, may be able to experience similar emotions evoked by that character. However, once "The Casting" was unveiled, the silence that fell upon the audience at E306 was more of shock and dismay than awe. The main character, Mary Smith, who was meant to be perceived as an empathetic, realistic, human-like character, failed to impress or engage people and was criticized for being uncanny. Not only was Mary Smith's skin pallor regarded as gray and dull, but also her facial expression was perceived as unnatural and wooden and failed to match the emotive qualities of her speech (Doerr,

2007; Gouskos, 2006). This female protagonist was intended to evoke pity from the audience for her misfortune, yet viewers failed to empathize with her and were put off by her strange and bizarre behavior. The audience was also aware of a distinct asynchrony of speech with her lip movement, which further exaggerated the uncanny for Mary Smith. Her performance was thus regarded as more comical than seriously thought-provoking. Quantic Dream returned to the drawing board, and on later release of *Heavy Rain* in 2010, despite claims that characters would portray the smallest details in facial expression, viewers were still left confused as to the affective state of (the intended to be) empathetic characters featured in the game (Crigger, 2010; Hoggins, 2010). Even though the graphical fidelity of characters such as the main protagonist, Ethan Mars, had improved, the audience could still detect imperfections in this character's facial expression and speech. The game designers meant for Ethan and other protagonist and antagonist characters to evoke heightened emotion in the player so that they may participate fully with the story (Doerr, 2007). However, awkward and strange facial expressions and a distinct lack of synchrony of lip movement with speech left some players uninspired and disengaged with the game (Hoggins, 2010).

Such feedback about characters featured in *Heavy Rain* provided some support that the Uncanny Valley phenomenon existed in realistic, human-like characters, yet this concept still required experimental investigation (Brenton et al., 2005; Pollick, 2010). Despite various attempts to simulate human-like facial expression and speech in animation and games (both in prerecorded and real-time footage), there is general agreement that largely such performances can fail to be perceived as real and suspend disbelief for the viewer. As such, the Uncanny Valley is now popular lexicon in the discourse of animation and games, and the phenomenon now occurs as a recognized design issue when creating realistic, human-like characters (see, e.g., Brenton et al., 2005; Doerr, 2007; Geller, 2008; Green et al., 2008; Ho and MacDorman, 2010; Hoggins, 2010; Plantec, 2007; Pollick, 2010; Van Someren Brand, 2011; Walker, 2009). Given the increasing importance of the effective communication of a character's emotional state to the viewer so that the viewer may engage with and understand the intentions of that character during the game (or watching a film), I chose to focus my research on how speech and facial expression of emotion may influence viewer perception of the uncanny (i.e., trigger experience of the Uncanny Valley effect) in virtual characters. Specifically, I recognized that, while facial expression and speech were reported as factors that may evoke the uncanny, there was no guidance (based on empirical evidence) for which particular aspects of facial expression and speech may exaggerate the uncanny and, importantly, how such factors may be used to control the uncanny in character design.

As described in this book, technological advancements in simulating realism in games and animation have increased at an exponential rate since 2006 when Mary Smith, arguably the harbinger of the Uncanny Valley in games (Walker, 2009), was introduced. As a result of this, various titles have been proclaimed as having finally overcome the Uncanny Valley. For example, the cinematic action-adventure game *The Last of Us* (Naughty Dog, 2013) has been critically acclaimed as a "masterpiece" and "one of the best games of its generation" (Game, 2013). Importantly, it has also been proclaimed as "having crossed the Uncanny Valley." However, the audience now approaches such claims with increasing caution, having been unconvinced in the past. As well as design considerations and how one may achieve successful expression choreography in human-like virtual characters, I consider the possible psychological triggers of the Uncanny Valley in characters with aberrant facial expression and the implications of this phenomenon in the real world. For example, what happens when realistic, human-like, virtual characters are used in applications designed for the purpose of education and assessment beyond that of entertainment? Furthermore, what are the possible repercussions for children when interacting with human-like virtual characters in animation and games for entertainment and education purposes? This book introduces several of my publications comprising a body of empirical and theoretical research that critically analyzes viewer perception of facial expression and qualities of speech in realistic, human-like video game characters and relates this to the Uncanny Valley. The aim of this book is to identify visual and auditory stimulus features that currently affect uncanniness in virtual characters and, ultimately (as part of an ongoing research project), to work toward a model based on a perceived lack of empathy in a virtual character due to a character's inauthentic physical characteristics. Furthermore, this model may help explain the possible psychological substrates of the uncanny and provide recommendations for how to control it in character design.

In the first chapter I provide a synopsis of twentieth-century psychological literature on the subject of the uncanny and how this concept was first introduced into contemporary thought. This includes the associations made by the German psychologist Ernst Jentsch (1906) on the subject of the uncanny with automata and the horror genre. Lucien Freud's

(1919) later psychoanalysis as to why we experience the uncanny is also discussed, including his explanations of why we may find some objects less aesthetically pleasing than others to the extent that they are disturbing and frightening. A description is then provided as to how, in the later stages of the twentieth century, the robot designer Masahiro Mori developed the notion of the Uncanny Valley (Mori, 1970/2012). Mori created a hypothetical graph to demonstrate the observations that he had made of viewer response toward robots with an increasing human-like appearance. Importantly, Mori's seminal work demonstrated that increased realism does not always imply increased acceptance on the part of the viewer and that this concept may be applied to human-like characters beyond the discourse of robotics. A critical examination of how the Uncanny Valley has been associated with near human-like characters featured in games and animation is also given in Chapter 1 to provide an overview of the public's response to uncanny characters in trade press and the animation and gaming community. As well as a closer inspection of the waxen, uncanny figures in the case studies, *The Casting* and *Heavy Rain*, further examples of games and animation that have been reported as having fallen into the Uncanny Valley are discussed, such as *Beowulf* (Zemeckis, 2007), *The Adventures of Tintin* (Spielberg, 2011) and the *Grand Theft Auto* (GTA) series including human-like characters in the latest *GTA V* (Rockstar North, 2013). Also, the implications of real-time versus prerecorded footage are explored, as are the different marker-based and markerless performance capture procedures behind augmented realism in games and animation.

Other researchers in the fields of robot–human interaction (RHI) and human–computer interaction (HCI) have also explored human perception of the uncanny when interacting with robotic and virtual characters. The next chapter provides a meta-analysis of this previous work and how these findings have helped move our understanding of the Uncanny Valley forward. Furthermore, recent theoretical works on the subject of the uncanny in realistic, human-like agents have helped inform my body of work on viewer perception of facial expression and speech in virtual characters. As Chapter 2 reveals, much of the previous work on the Uncanny Valley in synthetic agents had been conducted using still images as stimuli in experiments. However, this previous work offered limited information as to how a character's behavioral fidelity may exaggerate the Uncanny Valley. Accordingly, the following chapter describes a study that I conducted to investigate how cross-modal factors such as facial expression and speech may exaggerate viewer perception of the Uncanny Valley in animated virtual characters. Given that the uncanny is synonymous with the horror genre, the findings from this study also revealed how a character's facial expression and speech may be manipulated to mitigate the uncanny effect in empathetic characters or to *enhance* the uncanny for antipathetic characters, such as zombie characters intended to be strange and frightening featured in the horror genre. The results from this study are referred to throughout all chapters in this book as the *Uncanny Modality* study, unless otherwise stated from another study. The results from the Uncanny Modality study showed that characters were perceived to be more uncanny when there was a lack of human-likeness in the character's facial expression, particularly in the upper facial region, with the forehead and brows being of particular significance. Certain characteristics of a character's speech also made it appear stranger and less human-like, including speech that was of an incorrect pitch or tone and speech delivery that was regarded as slow, monotone or the wrong tempo. An overexaggeration of articulation of the mouth during speech increased the uncanny in characters, and speech that was not judged to be synchronized with mouth movement also made characters appear stranger. Characters that were rated as having close-to-perfect lip-synchronization were regarded as more familiar and less uncanny than those with an asynchrony of lip movement with speech. The implications of these findings in games and animation as well as how a designer may use such tactics to reduce or enhance the uncanny for empathetic or antipathetic characters are discussed.

The results of the Uncanny Modality study indicated that a perceived lack of human-likeness in facial expression, specifically above the eyelids, exaggerated the uncanny. In addition to body gestures and the spoken voice, facial expression is used to communicate how one is feeling and to interpret the affective state and possible actions of others (Darwin, 1872; Ekman, 1979, 1992a, 1992b, 2004; Ekman and Friesen, 1969, 1978). During speech, the upper face is predominantly used for nonverbal communication (NVC) because the lower face is restricted by the articulatory processes (Busso and Narayanan, 2006; Ekman, 1979, 2004; Ekman and Friesen, 1969, 1978). Given the physical restraints of the articulatory processes in the lower face and the noted importance of NVC in the upper face during speech, I wished to investigate the implications of aberrant facial expression in the upper and lower face for perception of the uncanny in virtual characters. Building on the importance of facial expression in humans as a means of communication and the findings of the Uncanny Modality study, I then designed two further experiments to examine more closely how a perceived lack of expressivity in the upper face region (including the eyelids, eyebrows and forehead) and an overexaggeration of mouth movement during speech may influence perception of the uncanny in characters. Chapter 4 describes the findings from these studies and the actual implications of this work in character design across different emotion types such as anger, disgust, fear, happiness, sadness and surprise.

Until now, the work that I and others have done on the subject of the uncanny in human-like synthetic agents has provided design guidelines as to how either to reduce the uncanny for characters intended to be regarded as empathetic or to exaggerate the uncanny in antipathetic characters intended to be perceived as frightening and disturbing. However, few have explored either the possible psychological drivers of the uncanny experience or the specific traits perceived in an uncanny character that may evoke such a negative response in the viewer. The findings from my initial experiments provided some evidence as to possible facial stimuli features that trigger experience of the uncanny in characters, which led me to question how aberrant facial expression in human-like virtual characters may be linked to possible psychological causes behind the Uncanny Valley. From Chapter 5 onward, I present an emerging theory as to the cause of the Uncanny Valley based on empirical evidence from my experiments and findings from other papers on human social interaction.

Many have defined one's ability to empathize with another and understand the cognitive and emotive processes of others as the unique quality that defines one as human (see, e.g., Azar, 2005; Blakeslee, 2006; Gallese et al., 1996; Hogan, 1969; Iacoboni et al., 2005; Winerman, 2005). I put forward that the cause of uncanniness in human-like virtual characters lies in the perception of a lack of empathy in that character. In this book, I discuss how the social brain interprets perceptual cues provided via facial expression and how a perceived lack of facial cues and an inability to decode vital nonverbal information may cause the Uncanny Valley effect. Importantly, the viewer may not only be unable to decipher that character's emotional state but also doubt as to whether that character can understand and empathize with how others are feeling. Accordingly, I present three possible (hypothetical) scenarios based on a perceived lack of NVC in a character (especially in the upper face) that would otherwise be used as a primary mechanism to interpret and decode the emotional state of another. The first of these—that perception of the uncanny may be due to a perception of possible antisocial tendencies in that character, verging on psychopathy—is considered in Chapter 5. In humans, a lack of a facial startle reflex in responding to frightening or shocking situations, including a widening of the eyes and raised brows, is a physiognomic marker for those diagnosed with a psychopathic condition. Therefore, characters that do not portray this NVC to others when appropriate during a game or animation may be regarded as unpredictable, with a cold personality and an impulsive nature and may be perceived as a possible threat.

Building on this theory that uncanniness may, at least in part, be caused by a perception of a lack of empathy in a character, I propose that a viewer may be concerned that the character cannot experience emotional empathy and understand the feelings (and thoughts) of the viewer. Furthermore, this raises alarm that the character may not be able to demonstrate compassion toward them. In Chapter 6, I consider that this assumption may be prompted when interacting with a character by a perceived lack of facial mimicry in a character to one's self and others. Recent findings in neuroscience on mirror neuron activity in humans are also discussed in this chapter to help support this theory as a possible psychological cause of the uncanny in human-like virtual characters. In Chapter 7, I consider how a viewer may interpret this perceived lack of facial mimicry in a character with the belief that there is an inability to forge a meaningful attachment with that character. Given that humans are acutely sensitive to and rely upon the acknowledgment of others to help justify one's own existence, a perceived lack of facial feedback from a character that they are expecting to behave as a human may result in abnegation (or doubt) of oneself. This state would be a highly destabilizing and uncomfortable experience for the viewer. Not only might it raise alarm that one is being ignored by a character, but also a realization may occur that this evident lack of response in a character may be due to possible antisocial traits within oneself. Rather than accept the possibility of abnormal social skills within oneself, the viewer finds fault with that character, hence collectively evoking the uncanny.

Finally, in Chapter 8, I provide a new theory in response to the common question posed by both the audience and character designers: Will we ever overcome the Uncanny Valley? Other researchers and authors suggest that, in time, this will be possible due to advancements in technology for simulating realism. However, based on the outcomes of my empirical work, I suggest that our discernment for detecting imperfections in a character's facial expression may keep pace with new developments in technology so that we are less likely to accept realistic, human-like virtual characters as authentic and believably human. This raises a unique and intriguing debate as to whether advances in technology will eventually allow us to cross the Uncanny Valley or whether new, human-like characters in games and animation are preordained to be regarded as uncanny despite improvements in simulating realism. As yet, there has been no book available for those in academia and industry to act as a reference to provide possible psychological explanations as to *why* the uncanny may occur in human-like characters and *how* this phenomenon may be controlled in character design. This book fills that gap in literature by assessing the biological and social roots of the uncanny and its implications for computer graphics animation and HCI.

REFERENCES

Azar, B. (2005) "How mimicry begat culture," *Monitor on Psychology*, vol. 36, no. 9, p. 54.

Blakeslee, S. (2006) "Cells that read minds," *New York Times*. Retrieved October 31, 2013, from http://www.nytimes.com/2006/01/10/science/10mirr.html.

Brenton, H., Gillies, M., Ballin, D. and Chatting, D. (2005) "The Uncanny Valley: Does it exist?" Paper presented at the HCI Group Annual Conference: Animated Characters Interaction Workshop, Napier University, Edinburgh, September 5–9.

Busso, C. and Narayanan, S. S. (2006) "Interplay between linguistic and affective goals in facial expression during emotional utterances," in *Proceedings of the 7th International Seminar on Speech Production*, Brazil, pp. 549–556.

Crigger, L. (2010) "Bad romance: Love in the time of videogames," *Gamespy*. Retrieved October 31, 2013, from http://www.gamespy.com/articles/108/1087383p2.html.

Darwin, C. (1872, reprinted 1965) *The Expression of the Emotions in Man and Animals*, Chicago: University of Chicago Press.

Doerr, N. (2007) "Heavy Rain devs have 'conquered' the Uncanny Valley," *Joystiq*. Retrieved October 31, 2013, from http://www.joystiq.com/2007/12/18/heavy-rain-devs-have-conquered-the-uncanny-valley/.

Ekman, P. (1979) "About brows: Emotional and conversational signals," in Von Cranach, M., Foppa, K., Lepenies, W. and Ploog, D. (eds.), *Human Ethology: Claims and Limits of a New Discipline*, New York: Cambridge University Press, pp. 169–202.

Ekman, P. (1992a) "An argument for basic emotions," *Cognition and Emotion*, vol. 6, pp. 169–200.

Ekman, P. (1992b) "Are there basic emotions?" *Psychological Review*, vol. 99, no. 3, pp. 550–553.

Ekman, P. (2004) "Emotional and conversational nonverbal signals," in Larrazabal, M. and Miranda, L. (eds.), *Language, Knowledge, and Representation*, Dordrecht, Netherlands: Kluwer Academic Publishers, pp. 39–50.

Ekman, P. and Friesen, W. V. (1969) "The repertoire of nonverbal behavior: Categories, origins, usage, and coding," *Semiotica*, vol. 1, pp. 49–98.

Ekman, P. and Friesen, W. V. (1978) *Facial Action Coding System: A Technique for the Measurement of Facial Movement*, Palo Alto, CA: Consulting Psychologists Press.

Freud, S. (1919) "The uncanny," in Strachey, J. (ed. and trans.), *The Standard Edition of the Complete Psychological Works of Sigmund Freud*, (1956–1974), London: Hogarth Press, vol. 17, pp. 217–256.

Gallese, V., Fadiga, L., Fogassi, L. and Rizzolatti, G. (1996) "Action recognition in the premotor cortex," *Brain*, vol. 119, no. 2, pp. 593–609.

Game (2013) "The last of us—Review," *Game*. Retrieved October 31, 2013, from http://www.game.co.uk/webapp/wcs/stores/servlet/ArticleView?articleId=232762&catalogId=10201&langId=44&storeId=10151.

Geller, T. (2008) "Overcoming the Uncanny Valley," *IEEE Computer Graphics and Applications*, vol. 28, no. 4, pp. 11–17.

Gouskos, C. (2006) "The depths of the Uncanny Valley," *Gamespot*. Retrieved October 31, 2013, from http://uk.gamespot.com/features/6153667/index.html.

Grand Theft Auto V (2013) [Computer game], Rockstar North (Developer), New York: Rockstar Games.

Green, R. D., MacDorman, K. F., Ho, C.-C. and Vasudevan, S. K. (2008) "Sensitivity to the proportions of faces that vary in human likeness," *Computers in Human Behavior*, vol. 24, no. 5, pp. 2456–2474.

Ho, C.-C. and MacDorman, K. F. (2010) "Revisiting the Uncanny Valley theory: Developing and validating an alternative to the Godspeed indices," *Computers in Human Behavior*, vol. 26, no. 6, pp. 1508–1518.

Hogan, R. (1969) "Development of an empathy scale," *Journal of Consulting and Clinical Psychology*, vol. 33, pp. 307–316.

Hoggins, T. (2010) "Heavy Rain video game review," *Telegraph*. Retrieved October 31, 2013, from http://www.telegraph.co.uk/technology/video-games/7196822/Heavy-Rain-video-game-review.html.

Iacoboni, M., Molnar-Szakacs, I., Gallese, V., Buccino, G., Mazziotta, J. C. and Rizzolatti, G. (2005) "Grasping the intentions of others with one's own mirror neuron system," *PLoS Biology*, vol. 3, pp. 529–535.

Jentsch, E. (1906) "On the psychology of the uncanny," *Angelaki* (trans. Sellars, R. 1997), vol. 2, no. 1, pp. 7–16.

Mori, M. (2012) "The uncanny valley," (MacDorman K. F. and Kageki, N. trans.), *IEEE Robotics and Automation*, vol. 19, no. 2, pp. 98–100. (Original work published 1970.)

Plantec, P. (2007) "Crossing the great Uncanny Valley," *Animation World Network*. Retrieved October 31, 2013, from http://www.awn.com/articles/production/crossing-great-uncanny-valley.

Pollick, F. E. (2010) "In search of the uncanny valley," *Lecture Notes of the Institute for Computer Sciences. Social Informatics and Telecommunications Engineering*, vol. 40, no. 4, pp. 69–78.

Ravaja, N., Turpeinen, M., Saari, T., Puttonen, S. and Keltikangas-Järvinen, L. (2008) "The psychophysiology of James Bond: Phasic emotional responses to violent video game events," *Emotion*, vol. 8, no. 1, pp. 114–120.

Sommerseth, H. (2007) "Gamic realism: Player perception, and action in video game play," in *Proceedings of Situated Play, DiGRA 2007 Conference*, Tokyo, Japan, pp. 765–768.

Spielberg, S. (producer/director) (2011) *The Adventures of Tintin: The Secret of the Unicorn* [Motion picture]. Los Angeles, CA: Paramount Pictures.

The Last of Us (2013) [Computer game], Naughty Dog (Developer), Japan: Sony Computer Entertainment.

Van Someren Brand, N. (2011) "Walking through the valley of life: Motion scanning as a bridge for crossing the Uncanny Valley," MA thesis, Utrecht University.

Walker, S. J. (2009) "A quick walk through uncanny valley," in Oddey, A. and White, C. A. (eds.), *Modes of Spectating*, Bristol, UK: Intellect Books, pp. 29–40.

Winerman, L. (2005) "The mind's mirror," *Monitor on Psychology*, vol. 36, no. 9, pp. 49–50.

Zemeckis, R. (producer/director) (2007) *Beowulf* [Motion picture]. Los Angeles, CA: ImageMovers.

CHAPTER 1

The Uncanny Valley

I START THIS JOURNEY BY examining what the uncanny is and what it stands for in relation to human-like virtual characters. Why should this old-fashioned adjective derived over a century ago be so pertinent in the most innovative CG character designs of today? On reading Jentsch's and Freud's psychoanalytical works, I acknowledged that the uncanny was indeed a feeling that I had experienced when I had encountered strange and scary objects. The word *uncanny* did describe the chill that I felt up my spine when I realized that there was something not quite right about a CG character presented with a near human-like appearance—to the extent that I was uncomfortable viewing the character's sinister depiction of a human-like form. Where did the root of this problem lie? Were there mistakes within the human-like character, or are we uncomfortable when presented with close simulations of the human form? In order to attempt to answer these questions, I had to understand the psychological drivers behind the uncanny and why Mori (1970/2012) made associations of this concept with robot design. In this first chapter I discuss rudimentary psychological literature on the subject of the uncanny and then why this phenomenon was recognized in robots with a human-like appearance. To demonstrate the public's response and critique of virtual characters with a human-like appearance, I include a section on the impact of the uncanny in games and animation in trade press. Building on this initial investigation into experience of the uncanny, in later chapters I tackle why we may be so intolerable of virtual characters with close human-likenesses, address problems in the production stages of character design and provide a new theory as to why factors such as abnormal facial expression and speech cause the Uncanny Valley in human-like virtual characters.

1.1 EXPERIENCE OF THE UNCANNY

Experience of the uncanny can be traced back to psychological writings of the early twentieth century when the psychologist Ernst Jentsch characterized the uncanny as a mental state that occurs when one cannot distinguish between what is imagined or real, or alive or dead. In his essay "On the Psychology of the Uncanny" (1906), Jentsch gave examples of lifelike wax dolls or automata as objects that may elicit an eerie sensation. He reasoned that such objects provoke *uncanniness* (i.e., experience of the uncanny) as the viewer cannot decide whether the object is real or unreal, or animate or inanimate. Furthermore, this effect may be exaggerated if viewing an object such as a life-size wax figure in near darkness, as this reduced lighting may hinder one's judgment as to whether the wax figure is actually human or not. Jentsch postulated that the uncanny is a disturbing and uncomfortable feeling that may escalate to a haunting, shocking and ghastly experience. Moreover, even after identifying that an object is an artificial man-made object, the unpleasant feelings evoked by the uncanny may still linger. Jenstch suggested that this intellectual uncertainty was the very essence of the uncanny and recognized that the concept of the uncanny is inextricably related to the horror genre. Specifically, characters featured in horror storytelling may evoke the uncanny if there is an uncertainty whether the character is human or not, thus inducing fear and suspense in the viewer. Jentsch stated that in horror storytelling, "… one of the most reliable artistic devices for producing uncanny effects easily is to leave the reader in uncertainty as to whether he has a human person or rather an automaton before him in the case of a particular character" (p. 13). Furthermore, he acknowledged that authors in the horror genre may use the uncanny effect to their advantage (even unwittingly so) to enhance the fear factor in their work. Jenstch gave the example of Ernst Theodor Amadeus Hoffmann as an author of fantasy/horror literature who had successfully exploited the disturbing feeling of tension created with an uncanny situation. Hoffman (1817) created an antagonist character named the Sandman, who instills terror in children as a cautionary tale if they misbehave. The Sandman steals the eyes from children who refuse to go to bed. This wicked character visits them and throws sand into their eyes so that their eyeballs will bleed and fall out. The Sandman's own children nest in the crescent moon, so he collects these naughty human children's eyes and takes them to feed to his own children, who have pointed, bird-like beaks. Even today, the Sandman is still used as a fable to warn children that he will visit them if they misbehave and that his part-bird, part-human children are waiting to be fed. Jentsch proposed that the Sandman and his children elicited the uncanny because the reader could not be sure if they were human or not. Accordingly, this strategy allowed Hoffman to establish a reputation as a virtuoso in horror literature due to such frightening encounters with fantastical, only part-human, characters.

Jentsch (1906) also put forward that this intellectual uncertainty may be provoked when encountering individuals diagnosed with a mental disorder whose behavior may be judged as peculiar or disturbing. If an individual steps outside the boundaries of what may be regarded as normal behavior due to mental illness and the situation may not be judged as ordinary or comical (e.g., in the case of one who is intoxicated by alcohol), then one may be wary of manifestations of insanity. Based on this, Jenstch suggested that the uncanny may occur on witnessing someone having an epileptic seizure, where an individual has lost control of his normal bodily behaviors. In such circumstances, it may appear that more supernatural or automated processes were at work, controlling that individual, due to the jerky movements of a fitting state. He was not in control of his thoughts or actions and may be at the mercy of a more powerful, dominating force in control of them. The uncanny occurs as one takes a defensive stance against such individuals (i.e., those perceived to be of an unstable mental psyche) as the viewer cannot predict how that person will be behave.

Jentsch's (1906) work on the uncanny is recognized as a precursor to Sigmund Freud's (1919) psychoanalysis on the subject of the uncanny named *Das Unheimliche* (*The Uncanny*). The uncanny intrigued Freud. He acknowledged that, in the field of the aesthetics, much previous work had been done to consider the qualities of feeling associated with what is beautiful; however, a full explanation as to why some objects appear unsettling or frightening to the viewer was still required. Freud described the uncanny as a unique, terrible sensation associated with all that is dreadful and abhorrent. "The subject of the 'uncanny' is a province of this kind. It is undoubtedly related to what is frightening—to what arouses dread and horror …" (p. 219). Importantly, Freud noted that even though most people may have experienced the uncanny at some point, for example, on encountering a human-like artificial doll or automaton, attempts to define what the uncanny was were still unclear. Freud wished to ascertain why the fearful and unpleasant feelings associated with the uncanny occur and what causes this phenomenon. To help achieve this, Freud first undertook a thorough inquiry as to the meaning of the word *heimlich* (translated as *canny* in English) as a way to help gain a better understanding of what its opposite, *unheimlich* (uncanny), actually stands for. Having consulted and considered dictionaries in various different languages, Freud provided definitions of the word *heimlich* as what is "familiar," "not strange" and "belonging to the house" (p. 222). Based on these definitions, Freud affirmed that the word *unheimlich* could stand for what is "*not* known" and "… the opposite of what is familiar" (p. 220). Therefore, an object may appear frightening and uncanny because it was not familiar or known to us. Indeed, Jentsch had earlier based his analogy of the uncanny on this primary definition of the word *unheimlich,* meaning the opposite of what is homely or familiar. In this way, Jentsch inferred that an uncanny object appears unhomely or unfamiliar, and the uncanny is felt when a person is not fully at ease with a situation. However, this direct definition of what is *unheimlich* is therefore unfamiliar did not fully satisfy Freud. As he argued, not all that is new and unfamiliar necessarily puts one ill at ease to the extent that one is frightened. Therefore, the full complexities of the uncanny were still yet to be revealed.

Freud (1919) postulated that the unfamiliar or undecided may not cause an object to appear strange or frightening but that the second, less common meaning of the word *heimlich* would be more appropriate when representing the uncanny. In addition to describing objects that are familiar to us, *heimlich* can also stand for what is hidden or obscured from view to prevent others from knowing about it, "on the one hand it means what is familiar and agreeable, and on the other, what is concealed and kept out of sight" (pp. 224–225). Freud rationalized that since *unheimlich* is commonly used as the opposite of the first meaning of *heimlich*, it may also correspond to the second meaning and represent everything that should have stayed secret and concealed but has now become visible. In this way, the uncanny stood for a revelation of what should have remained hidden or secret. Uncanniness occurred when objects or situations evoked a *sinister* revelation of what is normally concealed from human experience. In other words, we experience the uncanny when we identify something hideous or unsettling that we or others may have been attempting to hide. Furthermore, the uncanny exists as a revelation of the repressed that should remain out of sight (and one's conscious awareness) to maintain one's psychological health and emotional stability. This may be applicable to less positive traits or behaviors in others that have suddenly been revealed. Or, on a psychoanalytical level, it may refer to repressed thoughts in us that we abruptly remember. For example, we may have repressed a traumatic or negative thought or memory that we are suddenly reminded of due to our interaction with a strange, uncanny object. As Freud stated, our anxiety and fear associated with the uncanny is driven by "… something repressed which *recurs*" (p. 241).

Based on this reasoning, Freud (1919) suggested that this revelation of what should have remained hidden or secret may not necessarily be a novel or unfamiliar concept. Rather, it may be something once familiar to us that we had forgotten or repressed in our mind, and we are suddenly alerted to it. In other words, it is the supervening of an earlier unconscious event upon a later event that may seem strange yet familiar at the same time. Importantly, the uncanny exists not only as a revelation of what may be hidden in others but also what *may be repressed or hidden in oneself*. In this way, one is led to question not only whether an object is real or not but also the reality of one's sense of oneself, for example, one's personality traits, and how we may be perceived by others (or not) in the real world (Freud, 1919; Royle, 2011). Freud linked Jentsch's (1906) earlier observation that epilepsy in others may evoke the uncanny to his own insights and meaning of the uncanny in that it represents a revelation of the repressed. As such, if one perceives manifestations of insanity in another, then this may evoke a sinister revelation of the possibility of insanity or maladaptive behavior within oneself.

> The uncanny effect of epilepsy and of madness has the same origin. The layman sees in them the working of forces hitherto unsuspected in his fellow-men, but at the same time he is dimly aware of them in remote corners of his own being. (Freud, 1919, p. 243)

Based on this analogy, Freud summarized that the uncanny phenomenon does involve feelings of uncertainty, but primarily an uncertainty of oneself. This theory is explored further in later stages of this book as to how aberrant facial expression of emotion in human-like virtual characters may trigger doubt of who one is, how they may be perceived by others and a morbid anxiety of one's own existence.

Since these earlier theoretical works on the concept of the uncanny by Jentsch (1906) and Freud (1919), this phenomenon has been associated with a broad spectrum of topics including automata, literature, language, horror, psychology, aesthetics, the double, telepathy, dreams, manifestations of insanity and the uncanny as a reminder of death (Royle, 2011). Building on the associations made of the uncanny with automata, the next section explores how, in the latter half of the twentieth century, the uncanny was used to explain why viewers experienced an eerie sensation toward human-like android designs in the field of robotics.

1.2 *BUKIMI NO TANI*—THE UNCANNY VALLEY

The first industrial robotic arm was developed in 1959 by George Devol and Joseph Engelberger as a way to reduce employees and increase output and precision for their manufacturing company, Unimation (IFR, 2012). This mechanical arm was named the Unimate and was sublicensed as an industrial robotic arm to other companies at an international level (IFR, 2012). With a purely functional appearance and purpose, people who encountered industrial robots of this type were likely to show little affinity or rapport toward them. Just over a decade later, following this rapid increase of functional robots in the workplace and robotics in society (IFR, 2012), a robotics professor at the Tokyo Institute of Technology named Masahiro Mori had an insight about human perception toward robot design that he articulated in an essay called "The Uncanny Valley" (1970/2012). This essay was first published in 1970 in a Japanese journal called *Energy,* and Mori's words are as profound today as then.

Mori (1970/2012) was aware that some engineers were not satisfied with robots that had a mechanical appearance, which served a purely functional role, and wished to explore how robots may interact with humans in daily life. To do this, they were creating robots with a near human-like appearance using artificial materials for the robot's flesh, skin and hair to cover the mechanical parts. These human-like robots were commonly referred to as android designs, with the purpose of successfully mimicking and interacting with humans. However, Mori explained in his essay that he was skeptical of this ambitious drive to simulate (and even augment) human behavior. Importantly, he predicted a negative reaction to these android designs and that human–robot interaction relationships may not be as successful as the designers had hoped. This analogy was based on his observations that, despite technological advances in robotics for imitating humans, the androids failed to fully replicate the appearance and behavior of a human. As such, instead of the more positive (or neutral) response people experienced toward industrial robots with a mechanical appearance, people responded negatively to the androids—even to the point of revulsion.

To demonstrate this lack of affinity toward the android designs, Mori (1970/2012) drew a hypothetical graph that plotted perceived affinity (or rapport) toward a robot against how human-like the robot appeared (see Figure 1.1). Mori placed a mechanical robot at the first point in his plot, demonstrating the more passive level of affinity one may take toward a robot such as an industrial robotic arm with a nonhuman-like appearance. Further along the axis for human-likeness is a humanoid robot; while this robot may maintain a mechanical appearance, it has more human-like features such as a face, torso, limbs, hands and feet. Examples of humanoid robots can be found in toy robots, and Mori suggested that people (in particular children) may show an increased affinity toward humanoid robots as they may also exhibit human-like traits such as a face and smile. Mori stated that, even though a toy robot may have a more mechanical, sturdy appearance, children may be especially attracted to features such as its human-like face and shape.

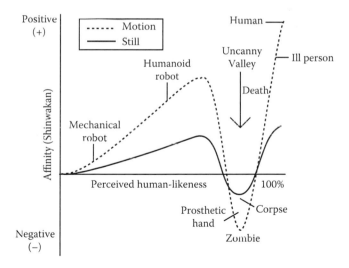

FIGURE 1.1 Mori's hypothetical graph of the Uncanny Valley that plots a viewer's perceived affinity toward an object against how human-like the object is. (Adapted with permission from Karl F. MacDorman and Norri Kageki's translation of "The Uncanny Valley" by Masahiro Mori, *IEEE Robotics and Automation* vol. 19, no. 2 (2012): pp. 98–100.)

The robot will start to have a roughly human-looking external form with a face, two arms, two legs, and a torso. Children seem to feel deeply attached to these toy robots. Hence, the toy robot is shown more than half-way up the first hill. (Mori, 1970/2012, p. 98)

Based on these more appealing human-like traits that children commonly respond to, Mori (1970/2012) justified why he had placed the toy robot (and other humanoid robots) above robots with a purely nonhuman-like mechanical appearance in terms of how familiar and likeable people would perceive them to be (see Figure 1.1). This increase in perceived affinity with human-likeness created a linear climb in Mori's diagram, yet it is not a continuous ascent. As shown in Mori's diagram, increased human-likeness in a robot's appearance is acceptable up to a particular point until, as with android designs, the robot appears close to human, but not fully. The increased human-likeness of an android's appearance elicits *increased* expectations in the viewer, and if the android does not meet these expectations then they risk being regarded as strange or creepy. In other words, we expect the android to behave as a human would based on their near human-like appearance, and if they fail to do so then we are alerted to a strangeness or difference about them. Intriguingly, if we realize that the android is not human because they do not accurately resemble a human in their appearance and/or behavior, this creates a lack of affinity toward the android to the extent of repulsion and we may experience an eerie, uncanny sensation. As Figure 1.1 shows, this sudden decrease in how accepting a viewer will be of an android causes a valley-shaped dip in the otherwise positive relationship between perceived affinity and human-likeness that Mori coined the *Uncanny Valley* (*bukimi no tani* in Japanese).

To help illustrate his theory, Mori (1970/2012) gave examples of objects such as zombies, corpses and life-like prosthetic hands that evoke a similar eerie, creepy sensation and lie at the nadir of the Uncanny Valley (see Figure 1.1). Based on these uncanny objects and the fact that movement is fundamental to humans, Mori predicted that the uncanny would be intensified with object movement. Mori used an example of a prosthetic hand to explain this increased viewer discomfort with movement and touch. Thematically, an electronic prosthetic hand may appear no different than a human hand. For example, the skin tone and fingers may resemble closely that of a human hand. However, when it moves, the jerky, mechanical movements of the hand joints may be shocking and thus may evoke a sense of the uncanny. Furthermore, touching the artificial hand may evoke alarm because of its hard, cold surface, which is different from the tactile qualities of soft, warm human skin (Mori 1970/2012).

In his essay, Mori (1970/2012) used the Japanese neologism *Shinwakan* to describe this unsettling feeling that he had experienced toward androids and other uncanny objects. This neologism that Mori originally used to measure the uncanny has more recently been translated in English as one's familiarity or affinity toward an object (Bartneck et al., 2009; Ho and MacDorman, 2010; see also Mori, 1970/2012). As a possible explanation of uncanniness, Mori (1970/2012) speculated that uncanny objects may remind us of the characteristics of a corpse as an inanimate, lifeless object. A viewer not only would be less likely to strike an affinity with and have a lower comfort level toward a corpse than a human but also would be reminded of his or her own death. In this way, experience of the uncanny occurs as the viewer is reminded that we are not immortal and that death is inevitable with the associated feelings of dread. Given the negative perceptual implications of uncanniness on human–robot interaction, Mori strongly advised that robot designers should avoid attempting to reach the second valley peak. Mori suggested that robot designers should think carefully about the effect their design may have on the viewer and how the robot they build will be received. To avoid the Uncanny Valley, robot designers should be motivated by increased affinity in their robots and not human-likeness. Rather than focusing on attaining near human-likeness, their goal should remain at the first valley peak to heighten affinity with the robot and prevent it tumbling into the valley.

> I recommend that designers instead take the first peak as their goal, which results in a moderate degree of human likeness and a considerable sense of affinity. In fact, I predict that it is possible to create a safe level of affinity by deliberately pursuing a nonhuman design. (Mori, 1970/2012, p. 100)

In addition to the field of robotics, the accelerating power of technology over the last decade in computer graphics (CG) has facilitated increased realism in gaming and animation with an increase of characters with a human-like appearance featured in these digital applications. Various film directors, animation studios and video game developers are going against Mori's recommendations and are pushing the boundaries of technology in the pursuit of realism. Despite Mori's warnings of the potential unease people may feel toward highly human-like androids, those working in games and animation continue to invest large amounts of time, money and resources into improving aesthetic realism for characters and environments with the intention to impress the audience. Yet, rather than improve their experience, in some cases the audience has been left disappointed by and put off an animation or game due to the eerie and creepy human-like characters they have encountered. As such, this has sparked a renewed curiosity in Mori's seminal theory of the Uncanny Valley. This provocative concept has evoked wide interest from both academia and the media, with numerous journal and conference papers and newspaper articles written on the subject of the Uncanny Valley in human-like virtual characters (and androids) over the last ten years. Furthermore, this interest should only intensify as technology continues to evolve to allow an increasing sophistication of realism in human-like characters. In the following section I discuss how a similar uncanny effect to that with androids has been reported in the trade press and in the gaming and animation communities when interacting with near human-like virtual characters in games and animation.

1.3 CRITICAL STUDIES OF THE UNCANNY VALLEY IN GAMES AND ANIMATION

Practitioners such as 3D animators for film and TV and games designers may have previously encountered the theory of the Uncanny Valley in scholarly articles such as journal papers or at academic conference presentations on the subject of the Uncanny Valley in CG animation. However, with increasing media attention to the Uncanny Valley phenomenon, they may also have read about this concept in trade press. This topic is also frequently discussed in the gaming and animation communities as an important (if not recurring) factor in new releases that have used realistic graphics.

By the early twenty-first century, motion capture techniques (commonly referred to as *mo-cap*) were frequently being used in cinematic film and animation, and this served as a catalyst for experience of the uncanny in characters with a human-like appearance. Mo-cap is a computer-based system that captures the movement of people and objects that is then transferred to a 3D model. Traditionally, motion-capture systems have used precision placed markers to capture movement from an actor's body or face. Actors wear suits with markers placed on their body and face, and their movements from these points are recorded by digital cameras surrounding them. The data collected from these points are then used to animate the same (or similar) points on a virtual character using animation software. More recently, markerless motion scanning techniques have been developed that rely on high-definition cameras to capture an actor's performance without having to wear markers on their body or face (Ballan and Cortelazzo, 2008; Freedman, 2012; Kehl and Gool, 2006). Either a single (monocular) or multiple (multiview) video cameras record the actor's movements, and these images are then processed to estimate the actor's position and pose at every point in time. Even though these methods offer a less invasive method so that an actor can move more freely without wearing special garments or devices, those researching the area of markerless mo-cap acknowledge that the reconstruction of the virtual character's movements (i.e., based on image cues) has its own challenges and restrictions (see, e.g., Ballan and Cortelazzo, 2008; Freedman, 2012; Kehl and Gool, 2006). First, estimating the bodily pose from video sequences can be difficult when the markers are omitted, as the link between observations and the actual pose is less clear. Second, capturing the accuracy and detail of motions at high speed or those of only a few millimeters (e.g., facial movements) may be limited by the camera's resolution and the sophistication of matching procedures between the predicted pose and the observed images (Kehl and Gool, 2006).

Media accounts of the Uncanny Valley in virtual characters first appeared following the release of the CG film *Final Fantasy: The Spirits Within,* directed by Hironobu Sakaguchi (2001). As this was the first mainstream CG film to be created fully using marker-based mo-cap and the first to include virtual characters with a human-like appearance, there was much anticipation as to the film's potential success and how the public may respond to these new, human-like virtual characters (Taylor, 2000). During the film's production, Sakaguchi (as quoted in Taylor, 2000) stated that this film would be revolutionary: "We've created characters that no longer feel blatantly computer generated …. It's something people have never seen before." The film did have a significant impact on the public, though not for the reasons that Sakaguchi may have intended. The main protagonist character, Doctor Aki Ross, was designed to be regarded as an empathetic, human-like character, yet her jerky movements and unnatural and emotionally limited facial expression evoked an adverse reaction from the audience. Instead of being perceived as attractive and likeable, she was regarded as creepy and strange. As Peter Plantec (2007) stated, she inadvertently came across to the viewer as unapproachable and rude due to her rather stoic, awkward demeanor, which was caused by anomalies in her facial expression and speech.

> [She is] a cartoon character masquerading as a human. As she moves, our minds pick up on the incorrectness. And as we focus on her eyes, mouth, skin and hair, they destroy the illusion of reality. Adding a voice we recognize (Ming-Na) only complicates matters. (p. 1)

Plantec grew so exasperated with Aki Ross that he described her as nothing more than a cartoon character *pretending* to be a more human-like character. Now, over a decade later, *Final Fantasy: The Spirits Within* is still regarded as an iconic example of the Uncanny Valley in the CG film industry (see, e.g., Beck et al., 2012).

Another film to rely solely on marker-based mo-cap was the children's animation *The Polar Express,* by Robert Zemeckis, which was shown in cinemas in late 2004. Oscar-winning actor Tom Hanks was represented as a human-like virtual character called the Conductor, who was intended to be perceived as an amiable character and appeal to a younger audience. Despite this character's friendly smile, which appeared on billboard advertisements and a humorous film narrative, attempts to establish a rapport with the audience were lost as soon as this character moved onscreen. The Conductor's motion was described as puppet-like (Jenkins, 2004), and the audience was critical of a lack of human-likeness in his facial expression that did not match the emotive qualities of his speech. The character's expressions also appeared out of context with a given situation as he presented an angry expression and a cold personality when interacting with other children characters in the film. Rather than a friendly, endearing protagonist, this character was reported as unfriendly with a stern demeanor (Jenkins, 2004). Provocative headlines such as "The Polar Express: A Virtual Train Wreck" (Jenkins, 2004) appeared, with much of this negative criticism attributed to the unsettling feeling that viewers experienced on watching Tom Hank's avatar. Based on this appraisal, *The Polar Express* is now renowned as a harbinger of the Uncanny Valley in children's animation (Beck et al., 2012; Jenkins, 2004; Pollick, 2010).

Despite this earlier unsuccessful attempt to generate a positive response to CG characters, Zemeckis decided to further exploit mo-cap technologies in his adventure sci-fi film *Beowulf,* which appeared in mainstream cinema in 2007. Vicon's marker-based mo-cap system was used to capture the facial and body movements of actors such as Angelina Jolie, featured as Grendel's mother, and Anthony Hopkins, as King Hrothgar. During this film's production, Zemeckis explained that the reason he chose to use mo-cap again for *Beowulf* was that it freed the actors (and characters) from the constraints that would normally be conflicted in live-action filming (Billington, 2007). Zemeckis elaborated that mo-cap allowed him to focus less on aspects such as lighting, the camera set-up and an actor's hair, makeup or costume and more on the actor's performance. Therefore, he anticipated that this process would improve that actor's performance and therefore the viewing experience. "The actors are liberated from the tyranny of a normal movie—it is absolute performance and great actors, like the ones in this movie, relish that" (Zemeckis as quoted in Billington, 2007). Yet rather than enhance the actors' performance, viewers were left disappointed at the lack of detail in the virtual characters' appearance and behavior. It appeared that this process had failed to capture the subtleties of an actor's gestures and facial expressions onscreen (Gallagher, 2007). As such, in close-ups of the characters' faces, the audience was left to fill in the blanks for this missing nonverbal information. During emotive scenes in the plot, characters that should have looked angry or fearful on the verge of death appeared unmoved with a perceived lack of intensity and authenticity in their facial expression (Gallagher, 2007). In this way, the film failed to suspend disbelief for the audience, and characters were perceived as odd and lifeless. The film critic Kenneth Turran (2007) suggested that Zemeckis was nothing more than a "willing slave to technological advances" for replacing "real" actors with disturbing digital avatars. In this CG film and those aforementioned, the actors' performance was compromised by the limitations of mo-cap technology with the audience being put off by the final outcome.

During the production of the CG film *The Adventures of Tintin: Secret of the Unicorn* (2011), Steven Spielberg worked closely with the animation team at Weta Digital to enable the actors' performance to be brought to virtual life using the latest mo-cap technologies. Spielberg attended the studios where filming took place so that he could personally direct the actors such as Jamie Bell, who played Tintin, and Andy Serkis, who played Captain Haddock, as they were filmed on a motion-capture stage. The actors wore head rigs as part of a new hardware and software facial-performance capture process that Weta Digital had developed for this film. In an interview with Barbara Robertson (2012), contributing editor for *Computer Graphics World*, Joe Letteri, a senior visual effects supervisor at Weta Digital, described this new facial mo-cap procedure. He said that much attention had been given to improving not only the visual and physical components of the skin texture also but the underlying muscle movement and physiognomy of the face. "We developed a new subsurface technique for the skin to have it look a little better, we developed some new facial software to add a layer of muscle simulation beyond what we could track and solve from the facial capture" (Letteri, as quoted in Robertson, 2012). The team at Weta Digital worked using the 3D software Autodesk Maya (commonly referred to as Maya) pipeline to create and sculpt the thousands of shapes used for the character's body and face. This allowed for the careful matching of the actor's observed poses with the virtual characters' movements also combined with key framed animation techniques to complete specific movements. The animators exploited Weta's bespoke plug-in software for Maya, called Tissue, to create dynamic movements in facial expression and further enhance details in the skin (Robertson, 2012). Deformations were applied to the existing face shapes when a particular shape is changed and reshaped. In this way deformation work was carried out to help create the appearance of muscle and fat beneath the skin. As such, details could be added such as folds and creases in the skin and wobbly cheeks as a character laughed.

However, even with this sophisticated facial animation software to add further detail to the characters facial expression, it became evident that acclaimed directors such as Spielberg and a team of up to sixty animators at Weta Digital (Robertson, 2012) were not able to escape the Uncanny Valley. Following the film's release in 2011, media reports stated that this near human-like representation of Tintin had fallen headfirst into the valley, and Tintin, rather than being a hero, was ridiculed as a hindrance to the film (Beifuss, 2011; Buchanan, 2011; Rose, 2011). Again, much attention was drawn to this character's abnormal facial expression as the audience grew frustrated at the evident lack of emotional expressivity from this character. Furthermore, many reported that there was a blatant disconnect between the heightened realism of the film sets with the more stylized realism of the characters (see, e.g., Beifuss, 2011; Buchanan, 2011). Kyle Buchanan (2011), of *New York Magazine,* observed that, while Tintin appeared likeable and charming, there was a mismatch between his behavioral fidelity with his realistic human-like appearance. "Tintin looks simultaneously too-human and not human at all, his face weirdly fetal, his eyes glassy and vacant instead of bursting with animated life." There was little difference between Tintin's calmer expression—when he sat, quietly contemplating a good day's work with his comrade, the little white terrier Scotty, or poring over sea maps for his next excursion—and his expression when he was being forced to jump from tall buildings and travel at high speed through busy Moroccan streets on a motorcycle in a fast-paced chase scene. Despite facing perilous scenarios in the movie, Tintin's face did not fully communicate fear, or at least excitement, in response to these high-drama scenes. As a result, some found him dull rather than intriguing and were left unconcerned if Tintin managed to escape from apparent danger or not. This was the case for John Beifuss (2011), who authored the article "'The Adventures of Tintin'—A Review: Steven Spielberg and the Uncanny Valley of Doom." He stated that despite being a film packed with action scenes, "we don't identify with him [Tintin] or fear for his safety, even when he's in danger or performing some spectacular feat." Overall, Tintin's glazed-over look and naive expression failed to match the sophisticated and highly impressive film sets and the animated, rich qualities of Jamie Bell's voice performance. Instead, Tintin's stare was judged as dead-eyed and distant, and the viewer was unable to engage effectively with this character.

In the same way that film directors such as Zemeckis and Spielberg have used mo-cap to push the boundaries of what can be achieved simulating realism in animation, game developers are seeking to enhance the player experience by implementing this technology in game aesthetics. David Cage, chief executive of the game development company Quantic Dream, has pioneered several projects with the intention to enable emotive cinematic drama in games. Cage's mission began with exploiting the capabilities of the Sony PlayStation 3 (PS3) (released to the public in 2006) to explore how this game engine may be used to convey human emotion. The theme of Cage's new game, *Heavy Rain* (Quantic Dream, 2010), was an interactive crime-thriller and was heavily influenced by the suspense and emotion evoked in film noir. The plot entailed solving the mystery of a serial killer, who continued to drown his victims in heavy rainfall until the player was able to identify the killer from clues in the game. In an interview with Jeremy Dunham from Imagine Games Network (IGN), Cage stated that he intended *Heavy Rain* not only to be innovative in its approach but also to provide the player with a novel, immersive and emotive experience. "We hope to break new grounds, explore new possibilities and create a very unique emotional experience" (Cage, as quoted in Dunham, 2006). However, Cage was also aware that the full impact of this psychological drama could not be realized without the characters' ability to portray convincing human emotion (Robinson, 2012). Another challenge that Cage faced was that the game's aesthetics (and characters' facial expressions) would run in real time from the PS3 console. In animation, footage is prerendered, and 3D animators may take advantage of the high pixel resolution achievable onscreen. Yet because game play occurs in real time, images are rendered in real time, thus relying on the power of the graphics chips and rendering capabilities of that game's console. As such, the level of realism achieved in gaming environments may be less than that achieved in animation due to technical limitations such as hardware resources and processing power (Ashcraft, 2008).

Even though gaming graphics lag behind that of the high polygon counts achieved for characters in animation, Cage continued his pursuit of realism with the belief that this would improve player engagement and immersion with this interactive medium. In an attempt to persuade the player that the virtual characters in his new game *Heavy Rain* (Quantic Dream, 2010) were believably human, Cage used Quantic Dream's in-house Vicon mo-cap system, including Vicon's 28 MX 40-foot high-definition (HD) cameras. It was hoped that this HD technology would provide facial capture data of the highest precision, including subtle nuances and the smallest details in the character's facial expressions. The world caught its first glimpse as to what had been achieved on the PS3 and what gaming experience *Heavy Rain* may offer the viewer when Cage unveiled a technical demonstration for the game at the E3 2006 entitled *The Casting*. The plot focused on a human-like character called Mary Smith, who provided a personal account as to the devastating impact of discovering her husband's affair. Mary's dramatic narrative was supposed to evoke sympathy from the audience, yet imperfections in her facial expression from the human norm evoked a lack of sympathy to the extent of repulsion (Doerr, 2007; Gouskos, 2006; Walker, 2009). Viewers were unable to empathize with Mary Smith as her facial expressions appeared insincere and did not match the sad, desperate tones in her voice. Furthermore, viewers found it difficult to discriminate between different emotions as one facial expression shifted awkwardly to the next (Doerr, 2007; Gouskos, 2006). On close inspection of her facial movements, Mary's inner eyebrows were not raised when she was supposed to be expressing sadness. While this is a subtler facial movement, it is crucial for the effective communication of sadness or to portray a more negative emotion (Ekman, 1979, 2004; Ekman and Friesen, 1978). Similarly, the corners of her mouth were often raised as if she was about to break into a smile rather than being pointed down toward her chin to convey the droopy mouth shape characteristic of experiencing sadness. The audience could hear that Mary was supposed to feel sadness from her words and qualities of speech, yet they were unable to detect evidence of this emotion in her face. This reduced not only the plausibility for this emotion but also the overall believability for this character, and she was regarded as insincere and strange. If Cage was to succeed in providing interactive human drama that would suspend disbelief via a gaming console, then the sophistication and authenticity of a character's emotion would need to be improved.

With Mary Smith having been placed at the nadir of the Uncanny Valley, Cage returned to the development studios at Quantic Dream to try to ensure that the characters in the final version of *Heavy Rain* would not suffer the same fate (Robinson, 2012). Three years later, the game was released to the public. However, despite a noticeable improvement in the graphical fidelity of the quality of the characters' skin, clothing and hair texture, still their behavioral fidelity did not match their human-like appearance. As had occurred with Mary Smith in the earlier prototype demonstration, there was a perception of a lack of human-likeness in the new characters' facial expression. A profound scene takes place as one of the main protagonist characters, Ethan Mars, loses his son in a busy shopping mall and searches panic-stricken for him. Ethan's purpose in the game is to protect his son from the serial killer, and it was important that players could empathize with Ethan in his plight so that they may engage fully with this scene. However, some players were critical of a distinct lack of believability in Ethan's facial expression as they were unable to understand his emotional state (Hoggins, 2010; Robinson, 2012). Players could detect the fear and panic in Ethan's voice as he shouted his son's name, but these emotions were not reflected in his facial expression. Ethan's lower face and mouth were fully animated as he spoke, but there was little movement in his upper face. As such, he failed to communicate the less obvious facial movements of fear such as raised outer eyelids and a furrowed, wrinkled brow (Ekman, 1979, 2004; Ekman and Friesen, 1978). Without this nonverbal information in the character's face, the player was unable to empathize with Ethan's emotional state and relate to his trepidation of being unable to find his son. As such, this scene (and others) fell below their expectations, thus preventing them from engaging with and participating in each progressing chapter (Hoggins, 2010). *Heavy Rain* was preempted to set new heights in human drama in games, yet a lack of human-likeness in the character's facial expression prevented some players from reaching this heightened emotive experience as they failed to empathize with the characters and engage with the plot (Dunham, 2006; Hoggins, 2010).

Humans have up to 43 facial muscles that can move independently of each other or in unison to create different expressions (Ekman and Friesen, 1978). As such, facial mo-cap can be more challenging than body motion capture due to the increased resolution required to capture and record only small movements from these muscles (Williams, 1990). For example, up to 350 markers may be applied to an actor's face in an attempt to detect more nuanced expressions from the eyes and lips. This issue is exaggerated in that comparisons have been made between inadequately animated, uncanny characters in video games and the typical affective reaction to unsuccessful cosmetic facial surgery in humans (Thompson, 2004, 2005). Marine Corps characters featured in the video game *Quake 4* (Raven Software, 2005) were reported as having an appearance similar to that of "victims of thoroughly botched face lifts" (Thompson, 2005, p. 1). Similarly, various female empathetic video game characters, such as Naomi from *King Kong* (Ubisoft, 2005), Sydney Bristow from *Alias* (Acclaim, 2004) and Serena St. Germaine from the James Bond video game series *Everything or Nothing* (Electronic Arts, 2004), have failed to achieve the desired viewer response, instead being described as "monsters" with "dead eyes" comparable to the zombie characters they are fighting (Thompson, 2004, 2005). Thompson (2004) made parallels between their stiff and awkward facial expression and the muscle-relaxing effects of Botox (Botulinum Toxin, Allergan, Inc., California): "Mouths and eyes don't move in synch. It's as if all the characters have been shot up with some ungodly amount of Botox and are no longer able to make Earthlike expressions" (p. 1). Despite their beautiful appearance and flawless skin texture, their aberrant facial expression repelled viewers. Instead of being regarded as heroines with whom the viewer could relate, viewers were exasperated by their lack of emotional expressivity.

There is continued unrest in the gaming community as to the quality and style of graphics and the Uncanny Valley phenomenon. Tron Knotts (2007) held a debate around the release of *Grand Theft Auto (GTA) IV* (Rockstar North, 2008) as to what constitutes "good graphics" and whether the pursuit of realism is a worthwhile commodity for the game. The GTA series boasts a mix of genres including action-adventure, driving and third-person shooting, and its urban environments and street life are inspired by American cities. Protagonists in the game aspire to defeat criminal underworld activities with various motives set in each GTA game based upon role-playing, using one's initiative to outwit antagonists and racing scenes. The first game was launched in 1997, and since then the developers have increased the visual realism with each release. Three-dimensional graphics were introduced in *GTA III* as a way to increase player immersion and to allow the player to explore and interact with 3D streets, vehicles and near human-like characters. However, using the GTA series as a case study, Knotts asked if continuously aiming to increase the number of pixels was sufficient to justify good graphics and a good gaming experience. Knotts compared the highly realistic graphic style used in *GTA IV* to closely simulate real-world characters and environments to a more stylized cel-shaded approach (used in earlier GTA games) to ascertain which was preferable and may be more comfortable for the player. Knotts clearly stated that his preference was to utilize the more basic, cel-shaded approach in GTA games. This rudimentary technique may be easier on the eye compared to its more unpredictable, realistic alternative, as characters that appear close to but not fully human can serve as an annoyance and distraction for players.

> *Grand Theft Auto IV* should employ the cel shaded look. The in-game graphics shown so far for the game are ugly as hell—far uglier than anything seen previously in the series. Sometimes too much detail and a near miss at realism is far worse to look at …. (Knotts, 2007, p. 1)

Again, as many others have observed, Knotts felt no greater sense of emotional contagion from the more realistic characters used in *GTA IV* than its more stylized predecessors, and he did not wish to connect with these "ugly" characters. Knotts reasoned that it is unlikely that video game graphics will be able to overcome the Uncanny Valley, at least not in the near future. As such, to prevent the player being disturbed by uncanny characters, implementing a reduced, more stylized realism in GTA and beyond can be just as appealing and visually interesting (if not better) for the player.

Rockstar Games set *GTA V* (Rockstar North, 2013) in a virtual city based on Los Angeles, California, and this time utilized its Rockstar Advanced Game Engine (RAGE) to present both automated and prerecorded performance capture footage that it hoped would enable the objects and characters to be perceived as authentically believable for players. However, one is reminded of Knotts's (2007) argument as one struggles with the continuous antinomic conflict of wanting to believe that the characters are real yet at the same time being aware of subtle flaws in their behavior and facial expression. One is aware (both in cut scenes and game play) that the upper face of the protagonist, Michael De Santa, appears out of unity with his lower face. This factor is emphasized further when he speaks, and it is as if his upper facial movements cannot keep pace with what he is saying to help communicate his thoughts and feelings. The human face communicates not only how someone is feeling but also processes unrelated to emotion such as mental effort, mental fatigue, concentrating on a given task, anticipation of events and startle reflexes (Boxtel, 2010; Veldhuizen, van Boxtel and Waterink, 1998). Without these understated facial movements to help communicate a person's inner conceptual, affective and physiological state, it is as if that person does not have the capacity to think, feel or respond to events around them. In other words, in the case of Michael De Santa and other characters in *GTA V*, it sometimes seems that the lights are on (this person is alive) but no one is home (there is something wrong with them).

The media can attract readers with sensational headlines such as "Tintin and the Dead-Eyed Zombies" (*Economist*, 2011), which focus on the derision and rejection of uncanny characters and the unsettling public reaction to games and animation. As such, much has been written about the negative affective response that an audience may take toward an uncanny near human-like virtual character (or robot), but few fully discuss the possible psychological drivers of experience of the uncanny, the specific physical traits in an uncanny character that may elicit this negative response or the possible beneficial impact of experiencing the uncanny in antagonist characters. The purpose of this book is to explore what can be learned from empirical research in games and animation studies along with ground-breaking findings in psychology and neurology and how this may help inform practical design decisions to control the Uncanny Valley in games and animation.

In his original diagram of the Uncanny Valley (see Figure 1.1), Mori (1970/2012) placed a human beyond the valley dip and postulated that for androids to escape the Uncanny Valley, first we must acquire a full understanding of what makes us human. "We should begin to build an accurate map of the uncanny valley so that through robotics research we can begin to understand what makes us human" (p. 100). To do this, designers must look beyond the discourse of design and to disciplines such as philosophy and psychology. Hence, the focus of this book looks beyond just aspects of character design and includes studies from philosophers and psychologists to better understand the essence of being human and apply this to character design. The importance of upper facial expression as a means of successful communication in humans is discussed further in this book, and novel explanations about why the Uncanny Valley occurs in animation and games are put forward. The anecdotal feedback reported in trade press and gaming and animation communities has provided some support that the Uncanny Valley phenomenon does exist in realistic, human-like characters. However, as other researchers have stated (see, e.g., Bartneck et al., 2009; Brenton et al., 2005; MacDorman and Ishiguro, 2006; Pollick, 2010), this concept still requires empirical validation. Accounts of empirical research undertaken in the field of robotics and animation and games to show that the Uncanny Valley does exist are discussed in the next chapter, along with the various reasons why we may experience the uncanny in synthetic human-like characters.

REFERENCES

Alias (2004) [Computer game], Acclaim (Developer), New York: Acclaim Entertainment.

Ashcraft, B. (2008) "How gaming is surpassing the uncanny valley," *Kotaku*. Retrieved November 1, 2013, from http://kotaku.com/5070250/how-gaming-is-surpassing-uncanny-valley.

Ballan, L. and Cortelazzo, G. M. (2008) "Marker-less motion capture of skinned models in a four camera set-up using optical flow and silhouettes," paper presented at *3DPVT'0—the Fourth International Symposium on 3D Data Processing, Visualization and Transmission*. Georgia Institute of Technology, Atlanta, GA, 18–20 June.

Bartneck, C., Kanda, T., Ishiguro, H. and Hagita, N. (2009) "My robotic doppelganger—A critical look at the Uncanny Valley theory," in *Proceedings of the 18th IEEE International Symposium on Robot and Human Interactive Communication*, Japan, pp. 269–276.

Beck, A., Stevens, B. K., Bard, A. and Cañamero, L. (2012) "Emotional body language displayed by artificial agents," *ACM Transactions on Interactive Intelligent Systems (TiiS—Special Issue on Affective Interaction in Natural Environments)*, vol. 2, no. 1, pp. 2–29.

Beifuss, J. (2011) "*The Adventures of Tintin*"—a review: Steven Spielberg and the uncanny valley of doom," *The Commercial Appeal*. Retrieved November 1, 2013, from http://blogs.commercialappeal.com/the_bloodshot_eye/2011/12/the-adventures-of-tintin-a-review.html.

Billington, A. (2007) "From motion capture to 3D: The technology of Beowulf," *Firstshowing.net Editorials*. Retrieved October 30, 2013, from http://www.firstshowing.net/2007/from-motion-capture-to-3d-the-technology-of-beowulf/.

Boxtel, A. van. (2010) "Facial EMG as a tool for inferring affective states," in *Proceedings of Measuring Behavior 2010*, pp. 104–108.

Brenton, H., Gillies, M., Ballin, D. and Chatting, D. (2005) "The Uncanny Valley: Does it exist?" paper presented at the *HCI Group Annual Conference: Animated Characters Interaction Workshop*, Napier University, Edinburgh, 5–9 September.

Buchanan, K. (2011) "The biggest problem with the Tintin movie might be Tintin himself," *New York Magazine*. Retrieved November 1, 2013, from http://www.vulture.com/2011/07/the_biggest_problem_with_the_t.html.

Doerr, N. (2007) "Heavy Rain devs have 'conquered' the Uncanny Valley," *Joystiq*. Retrieved October 31, 2013, from http://www.joystiq.com/2007/12/18/heavy-rain-devs-have-conquered-the-uncanny-valley/.

Dunham, J. (2006) "*Heavy Rain* interview," *Imagine Games Network*. Retrieved November 1, 2013, from http://www.ign.com/articles/2006/06/07/heavy-rain-interview?page=3.

Economist (2011) "Tintin and the dead-eyed zombies," *Economist Online*. Retrieved November 1, 2013, from http://www.economist.com/blogs/prospero/2011/10/performance-capture-animation#comments.

Ekman, P. (1979) "About brows: emotional and conversational signals," in Von Cranach, M., Foppa, K., Lepenies, W. and Ploog, D. (eds.), *Human Ethology: Claims and Limits of a New Discipline*, New York: Cambridge University Press, pp. 169–202.

Ekman, P. (2004) "Emotional and conversational nonverbal signals," in Larrazabal, M. and Miranda, L. (eds.), *Language, Knowledge, and Representation*, Dordrecht, Netherlands: Kluwer Academic Publishers, pp. 39–50.

Ekman, P. and Friesen, W. V. (1978) *Facial Action Coding System: A Technique for the Measurement of Facial Movement*, Palo Alto, CA: Consulting Psychologists Press.

Everything or Nothing (2004) [Computer game], Electronic Arts (Developer), Redwood City, CA: Electronic Arts Inc.

Freedman, Y. (2012) "Is it real … or is it motion capture? The battle to redefine animation in the age of digital performance," *Velvet Light Trap*, vol. 69, pp. 38–49. http://dx.doi.org/10.1353/vlt.2012.0001.

Freud, S. (1919) "The uncanny," in Strachey, J. and Freud, A. (eds.), *The Standard Edition of the Complete Psychological Works of Sigmund Freud*, London: Hogarth Press (1955), pp. 217–256.

Gallagher, D. F. (2007) "Digital actors in *Beowulf* are just uncanny," *New York Times*. Retrieved October 30, 2013, from http://bits.blogs.nytimes.com/2007/11/14/digital-actors-in-beowulf-are-just-uncanny/.

Gouskos, C. (2006) "The depths of the Uncanny Valley," *Gamespot*. Retrieved October 31, 2013, from http://uk.gamespot.com/features/6153667/index.html.

Grand Theft Auto IV (2008) [Computer game], Rockstar North (Developer), New York: Rockstar Games.

Grand Theft Auto V (2013) [Computer game], Rockstar North (Developer), New York: Rockstar Games.

Heavy Rain (2010) [Computer game], Quantic Dream (Developer), Japan: Sony Computer Entertainment.

Ho, C.-C. and MacDorman, K. F. (2010) "Revisiting the Uncanny Valley theory: Developing and validating an alternative to the Godspeed indices," *Computers in Human Behavior*, vol. 26, no. 6, pp. 1508–1518.

Hoffman, E. T. A. (1844) "The Sandman," in Oxenford, J. and Feiling, C. A. (eds.), *Tales from the German, Comprising Specimens from the Most Celebrated Authors*, New York: Harper & Brothers, pp. 140–164. (Original work published in 1817.)

Hoggins, T. (2010) "Heavy Rain video game review," *The Telegraph*. Retrieved October 31, 2013, from http://www.telegraph.co.uk/technology/video-games/7196822/Heavy-Rain-video-game-review.html.

International Federation of Robotics (IFR) (2012) "History of industrial robots: From the first installation until today." Retrieved October 31, 2013, from http://www.ifr.org/uploads/media/History_of_Industrial_Robots_online_brochure_by_IFR_2012.pdf.

Jenkins, W. (2004) "*The Polar Express*: A virtual train wreck," *The Ward-O-Matic*. Retrieved October 31, 2013, from http://wardomatic.blogspot.co.uk/2004/12/polar-express-virtual-train-wreck_18.html.

Jentsch, E. (1906) "On the psychology of the uncanny," *Angelaki* (trans. Sellars, R. 1997), vol. 2, no. 1, pp. 7–16.

Kehl, R. and Gool, L. V. (2006) "Markerless tracking of complex human motions from multiple views," *Computer Vision and Image Understanding*, vol. 104, pp. 190–209.

King Kong (2005) [Computer game], Ubisoft (Developer), Rennes, France: Ubisoft.

Knotts, T. (2007) "*GTA IV*, the uncanny V, and 'good graphics,'" *Destructoid*. Retrieved November 1, 2013, from http://www.destructoid.com/gta-iv-the-uncanny-v-and-good-graphics---39172.phtml.

MacDorman, K. F. and Ishiguro, H. (2006) "The uncanny advantage of using androids in cognitive and social science research," *Interaction Studies*, vol. 7, no. 3, pp. 297–337.

Mori, M. (2012) "The uncanny valley" (MacDorman, K. F. and Kageki, N., trans.), *IEEE Robotics and Automation*, vol. 19, no. 2, pp. 98–100. (Original work published in 1970.)

Plantec, P. (2007) "Crossing the great Uncanny Valley," *Animation World Network*. Retrieved October 31, 2013, from http://www.awn.com/articles/production/crossing-great-uncanny-valley.

Pollick, F. E. (2010) "In search of the uncanny valley," *Lecture Notes of the Institute for Computer Sciences. Social Informatics and Telecommunications Engineering*, vol. 40, no. 4, pp. 69–78.

Quake 4 (2005) [Computer game], Raven Software (Developer), Santa Monica, CA: Activision.

Grand Theft Auto IV (2008) [Computer game], Rockstar North (Developer), New York: Rockstar Games.

Robertson, B. (2012) "Animation evolution," *Computer Graphics World*, vol. 34, no. 9. Retrieved November 1, 2013, from http://www.cgw.com/Publications/CGW/2011/Volume-34-Issue-9-Dec-Jan-2012-/Animation-Evolution.aspx.

Robinson, M. (2012) "Introducing Quantic Dream's Kara," Eurogamer. net. Retrieved August 29, 2014, from http://www.eurogamer.net/articles/2012-03-07-introducing-quantic-dreams-kara.

Rose, S. (2011) "Tintin and the uncanny valley: When CGI gets too real," *Guardian*. Retrieved November 1, 2013, from http://www.guardian.co.uk/film/2011/oct/27/tintin-uncanny-valley-computer-graphics.

Royle, N. (2011) *The Uncanny*, Manchester, UK: Manchester University Press.

Sakaguchi, H. (producer/director) (2001) *Final Fantasy: The Spirits Within* [Motion picture]. Culver City, CA: Columbia Pictures.

Spielberg, S. (producer/director) (2011) *The Adventures of Tintin: The Secret of the Unicorn* [Motion picture]. Los Angeles, CA: Paramount Pictures.

Taylor, C. (2000) "Cinema: A painstaking fantasy," *Time*. Retrieved October 31, 2013, from http://www.time.com/time/magazine/article/0,9171,997597,00.html.

Thompson, C. (2004) "The undead zone," *Slate*. Retrieved November 1, 2013, from http://www.slate.com/articles/technology/gaming/2004/06/the_undead_zone.html.

Thompson, C. (2005) "Monsters of photorealism," *Wired*. Retrieved November 1, 2013, from http://archive.wired.com/gaming/gamingreviews/commentary/games/2005/12/69739.

Turran, K. (2007) "*Beowulf* sexes up, dumbs down an epic," *NPR*. Retrieved October 31, 2013, from http://www.npr.org/templates/story/story.php?storyId=16334109.

Veldhuizen, I. J. T., van Boxtel, A. and Waterink, W. (1998) "Tonic facial EMG activity during sustained information processing: Effects of task load and achievement motivation," *Journal of Psychophysiology*, vol. 12, pp. 188–189.

Walker, S. J. (2009) "A quick walk through uncanny valley," in Oddey, A. and White, C. A. (eds.), *Modes of Spectating*, Bristol, UK: Intellect Books, pp. 29–40.

Williams, L. (1990) "Performance-driven facial animation," in *Proceedings of the 17th annual conference on Computer graphics and interactive techniques, ACM SIGGRAPH Computer Graphics*, vol. 24, no. 4, pp. 235–242.

Zemeckis, R. (2004) *The Polar Express* [Motion picture]. Los Angeles, CA: Castle Rock Entertainment.

Zemeckis, R. (2007). *Beowulf* [Motion picture]. Los Angeles, CA: ImageMovers.

CHAPTER 2

Previous Investigation into the Uncanny Valley

R ESEARCH ON VIEWER PERCEPTION of human-like agents in the context of the Uncanny Valley had laid dormant since Masahiro Mori's seminal theory in 1970, but the increase of human-like characters in animation and games has sparked a renewed curiosity in this phenomenon (Green et al., 2008; Pollick, 2010; Steckenfinger and Ghazanfar, 2009). In addition to the articles and opinion expressed by those in trade press and the animation and gaming community, researchers have responded to this concept to investigate if the Uncanny Valley does exist in synthetic agents with a near human-like appearance. In a recent interview with Norri Kageki, a journalist and acting manager of the robotics division at Kawada Industries, Mori (1970/2012) stated that the Uncanny Valley applied to a wide range of disciplines beyond just that of design. "The Uncanny Valley relates to many disciplines, including philosophy, psychology, and design, and that is why I think it has generated so much interest"(p. 1). Certainly, interest in the Uncanny Valley has gained momentum appealing to researchers in domains as varied as computer games, film, psychology, philosophy, sound studies, medicine, education, neurology and the military, to name but a few.

Hence, to establish where research has led to so far on the Uncanny Valley and to set the new emerging theories I present in this book in context, this chapter provides an overview of previous theoretical and empirical studies conducted on the Uncanny Valley in synthetic agents.

While it does not present an exhaustive account of all studies exploring the Uncanny Valley, the synopsis of previous research will include studies undertaken on aspects of design such as a character's appearance and movement. Previous attempts to plot an Uncanny Valley with robots and virtual characters and how these relate to Mori's original hypothetical diagram are explored. Potential limitations in Mori's theory are addressed, on the grounds that some researchers have suggested that the actual meaning of the phenomenon is prone to ambiguity and may have been lost in translation. The effect of people's age and gender on sensitivity to the uncanny are also discussed. Finally, the question of nature or nurture is raised, as researchers have explored whether we may be born with an innate response to the uncanny or if it is developmental and relies on factors such as effective social interaction from infancy.

2.1 DESIGN GUIDELINES FOR A CHARACTER'S APPEARANCE

Various design guidelines have been authored to advise designers on how to prevent uncanniness in the appearance of human-like synthetic agents. For example, Robert Green and colleagues (2008) designed an experiment using images of humans, androids, mechanical-looking robots and two- and three-dimensional human-like virtual characters to investigate how aspects of facial proportion may be manipulated to control the uncanny. The main focus of the study was to assess how deviating from the best and most attractive facial proportions identified in humans may affect uncanniness for the perceiver. Male and female adults rated still images of headshots of mechanical and humanoid robots, male and female humans and human-like virtual characters of both genders ranging from 2D cartoon-like to a more 3D realistic appearance. The results showed that overall participants preferred narrower jaws and cheeks in characters. While wider-set eyes were favored in robotic characters, a narrower set eye was preferable in humans and human-like virtual characters. Importantly, it was suggested that the uncanny may be reduced (i.e., the character would appear more attractive and human-like) by increasing the size of the eyes, lips, and length of the face in a character (Green et al., 2008). However, it was suggested that designers should implement any manipulations in a character's facial proportion with caution, and, as Jun'ichiro Seyama and Ruth Nagayama (2007) found, a high degree of abnormality in a character's facial features exaggerated the uncanny. For example scaling the eyes to 150 percent of their original size may make a human-like character appear less realistic and more uncanny.

Humans use the technique of anthropomorphism to attribute human-like traits to inanimate objects and animals (Duffy, 2003). This allows us to rationalize a character's actions and accept them as a social companion, despite them being a nonhuman entity. For example, in the same way that we may attribute human-like characteristics and a personality to one's pet dog or cat, we can assign cognitive (thoughts) and emotional states to a nonhuman-like object (Duffy, 2003). As Mori (1970/2012) predicted, empirical research has identified that anthropomorphic character designs that lack a nonhuman appearance but that portray human-like traits are regarded more positively than realistic, human-like characters (Schneider, Wang and Yang, 2007). In 2007, Edward Schneider and colleagues conducted an experiment in which 60 participants rated still images of virtual characters, including characters from Japanese animation and video games, to test the potential relationship between perceived human-likeness and levels of attraction toward a character. The characters' appearance varied in the experiment, including those with a low level of human-likeness, such as a one-eyed cat named Kidrobot, the fish character Nemo from *Finding Nemo* (Stanton, 2003) and Tweety Bird from the *Looney Tunes* cartoons. Antagonist characters were also used as stimuli, for example an ogre from the massively multiplayer online game (MMOG) *World of Warcraft* (Blizzard, 2004), in addition to more stylized human-like characters such as Megaman (Capcom, 1987–2013), Mario from *Super Mario Sunshine* (Nintendo EAD, 2002) and Lara Croft from the 2006 release of *Lara Croft Tomb Raider: Legend* (Crystal Dynamics). The results showed that antagonist characters like the ogre from *World of Warcraft* and characters with a more realistic, human-like appearance achieved the lowest attraction scores. Characters with a more stylized human-like appearance, such as Mario and Lara Croft, were rated as most attractive and human-like. However, many characters with a more human-like appearance were not regarded so favorably. The authors concluded that, based on their results, increasing the human-likeness for a character does not always improve that character nor guarantee increased attraction for that character. Indeed, they stated, "The safest combination for a character designer seems to be a clearly non human appearance with the ability to emote like a human" (Schneider et al., 2007, p. 548). When images of animal characters such as Tweety Bird, Bugs Bunny or Snoopy were presented with a human-like smile and/or human-like hands, these human-like traits significantly increased a participant's affinity and attraction toward that character (Schneider et al., 2007). This was due to a participant's ability to project human-like traits such as thoughts and feelings onto the anthropomorphic characters due to their human-like characteristics.

As Mori (1970/2012) acknowledged, our ability to anthropomorphize nonhuman-like objects is paramount in the context of the Uncanny Valley. This is demonstrated in a study by the robot designer David Hanson (2006), who found that it was possible to design against the uncanny by giving human-like characters a less realistic and a more stylized, cartoonish appearance. Hanson used a morphing technique to present a set of images across a spectrum of human-likeness that morphed from human to android on one-half of the spectrum and android to a mechanical, humanoid robot on the other half. The results depicted an uncanny region in images of the android and morphs between the android and mechanical, humanoid robot that were rated as more eerie and less appealing than other images across the spectrum. However, Hanson made adjustments to the facial features of these uncanny characters to give them a more cartoon-like and friendly appearance. Due to these changes, when tested again the uncanny was less noticeable (almost eradicated) in the android and android/mechanical robot morphs where the uncanny had previously existed (Hanson, 2006).

Only still images were used as stimuli in these previous experiments, and this provided a somewhat limited investigation of uncanniness based on unresponsive, still images. Most characters featured in animation and games are not stationary, with motion, timing, dynamic facial expression and speech all being important aspects that may exaggerate the uncanny (Richards, 2008; Weschler, 2002). All of these previous authors have acknowledged that their results may have differed had they used animated characters with sound (Green et al., 2008; Hanson, 2006; Schneider et al., 2007; Seyama and Nagayama, 2007). While these studies do provide useful information with regards to a character's visual appearance, much more work is required to investigate a character's behavioral fidelity. The next section explores which aspects of a character's movement and behavior have been identified as exaggerating perception of the uncanny.

2.2 THE EFFECT OF MOVEMENT

The complexities of the Uncanny Valley are further revealed in previous studies undertaken in this field with animated synthetic agents, including androids and virtual characters. Viewers expect that, when presented with a human-like synthetic agent, the behavior of the android (Bartneck et al., 2009; Ho, MacDorman and Pramono, 2008) or virtual character (Vinayagamoorthy, Steed and Slater, 2005) will match their human-like appearance: any deviances from the human norm in sound and motion will alert the viewer to a sense of strangeness in that character. In other words, a perceived mismatch in the character's appearance and behavior exaggerates the uncanny. Researchers in the field of robotics, such as Karl F. MacDorman at the Indiana University School of Informatics, have referred to the behavior of a realistic android perceived as natural and appropriate when engaging with a human as "contingent interaction" (Ho et al., 2008, p. 170). One may expect an android to smile or nod to recognize one's presence or to look directly at the person with whom they are interacting, as they would typically expect from a human. If a person puts out a hand to initiate a handshake with the android, the android would be expected to reciprocate with a handshake. Hence, this contingent interaction is a key factor in assessing a human's response to an android (Bartneck et al., 2009; Ho et al., 2008; Kanda et al., 2004; MacDorman and Ishiguro, 2006). When contingent interaction is lacking and there is a perceived incongruence in the gesture and timing of an android's movements or an inappropriate response from the android to others and external events, then this has been found to exaggerate viewer perception of the uncanny (Ho et al., 2008; Minato et al., 2004; MacDorman and Ishiguro, 2006; MacDorman et al., 2010). For example, if an android did not lift its hand to shake the human's outstretched hand or if its movement is jerky and awkward, rather than smooth and controlled, then viewers may be put off by the android. They may perceive the android as rude for not attempting to reciprocate this friendly gesture or may be alerted to the fact that the android is nothing more than a mechanical, synthetic object due to its unnatural, jerky motion. Either way, such an experience may negate viewers' experience so that they avoid future interaction with the uncanny android.

Similarities have been revealed in viewer perception of the quality of body movements in androids and human-like virtual characters. Specifically, smooth motion is always preferred to more abrupt, uncontrolled (to the extent of degradation) body movements (see, e.g., MacDorman et al., 2010; White, McKay and Pollick, 2007). Furthermore (and as I focus on in this book), viewer awareness of congruent interaction also extends to realistic, human-like virtual characters (Hodgins et al., 2010; Vinayagamoorthy et al., 2005). Whether it is one-on-one interaction with a human-like conversational agent or observing how a human-like virtual character responds to other characters or situations in an animation or game, incongruence in their behavior and actions with what may otherwise be expected exaggerate the uncanny for the perceiver. In a study undertaken by Vinoba Vinayagamoorthy and colleagues (2005) to assess how a character's appearance and behavior may affect user experience in a virtual environment, it was reported that unless the behavioral fidelity of a character matched their visual fidelity and the context of the virtual environment in which the character was placed, then this may impede on user experience not only with that character, but with the virtual environment itself. As Vinayagamoorthy et al. stated,

> It is important to maintain consistent levels of behavioral fidelity with increasing levels of visual realism … The behavioral expressivity of a character must be modeled in correlation to the context within which the character is placed. The right level of subtle cues can perform surprisingly well …. (p. 124)

Vinayagamoorthy et al. (2005) recognized the importance of subtler nonverbal communication such as eye gaze during face-to-face communication with a human-like virtual character. In an experiment to assess the quality of communication with a human-like virtual character, Vinayagamoorthy, Steed, and Slater (2004) mapped the same realistic eye-gaze model onto a human-like character with a cartoon-like appearance and a human-like character with a photo-realistic appearance. Interestingly, participants in the study reported that applying the realistic eye-gaze model to the human-like character with a more cartoonish appearance did not improve their perception of the quality of communication with this character. However, the participants reported that they felt an improved quality of communication with the more realistic, human-like character that had realistic eye-gaze behavior. In this way, maintaining appropriate eye contact (i.e., contingent interaction) when communicating with the realistic, human-like character was of greater importance than when participants communicated with the cartoonish character. The participants expected *more* from the realistic, human-like character due to its increased human-like appearance, yet increasing the behavioral fidelity and authenticity on the part of the cartoon-like character did not improve participants' experience with it. Vinayagamoorthy et al.'s work showed that we do have a willingness to interact with virtual characters of a more sophisticated human-like appearance, yet as Mori (1970/2012) predicted, we are more discerning of characters with a near human-like appearance than those with a reduced graphical fidelity in human-likeness.

2.3 PLOTTING THE UNCANNY VALLEY

Given the complexities of simulating human movement and the different contexts within which a perceiver may regard a human-like synthetic agent, rather than just one Uncanny Valley previous authors have predicted that it may be more complex than the simplistic valley shape Mori (1970/2012) originally plotted. In other words, there may be multiple valleys with a plethora of factors contributing to the valley dips, especially so with animated characters (Bartneck et al., 2009; Hanson, 2006; MacDorman, 2006; Pollick, 2010; Tinwell and Grimshaw, 2009). Indeed, those who have attempted to plot an Uncanny Valley in which participant ratings of perceived uncanniness were scored for synthetic agents presented over a continuum of human-likeness support this prediction (e.g., see experiments by Bartneck, et al., 2009; Hanson, 2006; MacDorman, 2006; Tinwell, 2009; Tinwell and Grimshaw, 2009). The results from these studies do not comply with the two-dimensional construct that Mori envisaged and instead suggest a multidimensional model of Uncanny Valley. When Karl F. MacDorman (2006) plotted scores for videos of robots ranging from a mechanical appearance to a human-like appearance (which included some robots with speech) for perceived familiarity against human-likeness, the plot was inconsistent with Mori's Uncanny Valley. First, there was no significant, deep valley shape similar to the depth and gradient of Mori's diagram, and second, the results showed that robots with the same score for human-likeness could achieve significantly different scores for perceived familiarity.

Intrigued by MacDorman's (2006) findings with robots and seeking to find empirical evidence that the Uncanny Valley did exist in human-like virtual characters, I conducted an experiment titled "Uncanny as Usability Obstacle" using videos of 12 animated virtual characters and one human. Some of the virtual characters had speech and other vocalization, but not all (Tinwell, 2009). In response to the anecdotal reports of the uncanny in human-like video game characters, as well as to attempt to plot the Uncanny Valley, I wanted to investigate how an uncanny character may affect a player's overall satisfaction with a game. Therefore, in addition to asking the 65 participants to rate characters for perceived familiarity and human-likeness, I also asked participants to rate how satisfactory they perceived a character to be within the context of a video game and how much they would enjoy interacting with that character within a video game. The Web-based questionnaire was designed so that the participants watched the video of a particular character and then scored the following on a nine-point scale: how human-like they perceived the character to be (1 = nonhuman-like to 9 = very human-like); how strange or familiar they found the character to be (1 = very strange to 9 = very familiar); and how satisfactory they judged the character to be (1 = dissatisfactory to 9 = very satisfactory). The 12 characters in the study were as follows: five human-like characters—Emily and the Warrior by Image Metrics (2008); Mary Smith from Quantic Dream's (2006) tech demo *The Casting*; Alex Shepherd from *Silent Hill Homecoming* (Konami, 2003); and Brucie Kibbutz from *Grand Theft Auto IV* (Rockstar North, 2008); two zombie characters— one from *Silent Hill Homecoming* (Konami, 2003) and another from *Alone in the Dark* (Atari Interactive, Inc., 2009); three stylized human-like characters—a Chatbot, Lara Croft and Nintendo's Super Mario; and two characters with an anthropomorphic appearance—a Sackboy from *LittleBigPlanet* (Sony, 2008) and Sonic the Hedgehog. The virtual characters and human were placed in different settings and engaged in different activities in the videos.[*] An example of a male character with a realistic, human-like appearance, typical of those used in games and animation, is provided in Figure 2.1.

When character average ratings for perceived familiarity (on the y axis) were plotted against perceived human-likeness (on the x axis), rather than just one valley shape there were multiple valleys. As others had predicted (Bartneck et al., 2009; Hanson, 2006; MacDorman, 2006; Pollick, 2010; Tinwell and Grimshaw, 2009), the plot was much more complex than Mori's smooth curve and the valley dips were less steep than Mori's (1970/2012) perpendicular climb. The most significant valley was positioned at about 40–50% human-likeness, which was lower than the 80–85% human-likeness that Mori predicted with the character Super Mario on the left and the stylized, more human-like character Lara Croft on the right. Another valley nadir also occurred at about 70–80% human-likeness, leading from Lara Croft on the left to a human on the right. However, this was shallower than the other valley shapes. The results of my initial experiment did support some of Mori's original predictions in his theory of the Uncanny Valley. Anthropomorphic characters such as Sonic the

---

[*] Full details of the experiment design and results can be found in Tinwell (2009).

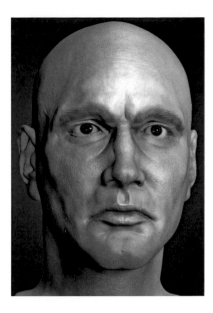

FIGURE 2.1 An example of a male, human-like virtual character by 3D artist Lance Wilkinson.

Hedgehog and Super Mario were placed in the top left corner of the graph, before a valley dip, and were regarded more favorably than other characters despite their low ratings for human-likeness (Tinwell, 2009). Conversely, the human was placed to the far right of the graph, being rated as most human-like and thus escaping the valleys. Furthermore, those characters with a perceived lack of behavioral fidelity in keeping with their human-like appearance, such as Mary Smith, were confined to the depths of a valley dip, along with the creepy zombie characters. As MacDorman (2006) found in his earlier study with robots, characters scoring similar ratings for human-likeness were rated very differently for perceived familiarity. In my experiment, a zombie character was placed in the bottom left-hand corner of the graph, indicating the lowest scores for perceived familiarity (i.e., rated as *most strange*) but sharing a similar human-likeness rating to Sonic the Hedgehog, who scored one of the highest ratings for perceived familiarity. This left me with an important question: Why did this occur? The scores and feedback provided for these characters with regards to user satisfaction in the context of a video game started to provide me with some indication of other factors that may contribute to uncanniness.

When the average scores for how satisfactory the 12 characters and human were regarded in the context of a video game were plotted (on the y axis) against average scores for perceived familiarity (on the x axis), the results revealed a strong relationship between a user's satisfaction score and how strange or familiar a character was perceived to be. Overall, the strangest characters (i.e., those rated as *least* familiar), such as Mary Smith, were also rated as the least satisfactory. In this way, I proposed that uncanny, human-like virtual characters may serve as a usability obstacle in a video game as the user is dissatisfied with that character. Hence, a perception of the uncanny in an empathetic human-like character not intended to be regarded as strange may have an adverse effect on a player's overall satisfaction and enjoyment with that video game (Tinwell, 2009). The three iconic characters—Lara Croft, Mario and Sonic the Hedgehog— were rated as highly familiar and much more satisfactory than characters with a more realistic human-like appearance such as Brucie Kibbutz, Alex Shepherd and Emily. However, the Sackboy was rated as less satisfactory than Sonic the Hedgehog, which suggests that the user demands anthropomorphism of a higher sophistication than the more simplistic appearance, behavior and limited human-like traits of the Sackboy character.

In keeping with the anecdotal criticism and negative attention that Mary Smith had received, this character was rated on a similar satisfaction level (and strangeness rating) as a zombie character. This was the first empirical evidence to show that the uncanny phenomenon was palpable for this human-like character, as she provoked a similar response in participants to that of an antipathetic horror character (Tinwell, 2009). Not only was the stylized Chatbot character perceived as just as strange as a zombie character but also she was rated as the most unsatisfactory character within the context of a video game. Feedback provided from some of the participants during the survey showed that this character's jerky motion, limited facial expression and poor quality of speech resulted in unusual dulcet tones that exaggerated how strange and unsatisfactory it was judged to be. Participants described this character as irritating and firmly disliked it. The findings from this initial study lead me to conclude that players do expect more than the limited and poor quality behavioral fidelity of characters such as the Chatbot within the context of a video game (Tinwell, 2009). Furthermore, given that overall the video of a human was perceived as less satisfactory than 3D virtual characters within the context of a video game, the pursuit for realistic human-like characters may be worth the endeavor. However, until video game developers have access to technology and expertise sufficient to overcome the Uncanny Valley, characters with a stylized or anthropomorphic appearance may be more satisfactory for players as they are less prone to uncanniness (Tinwell, 2009).

The results from experiments conducted with robots (MacDorman, 2006) and my study with virtual characters with motion and sound (Tinwell, 2009) demonstrated that the uncanny phenomenon is unlikely to be reduced to just two factors: perceived familiarity and human-likeness. While my study had gone some way in confirming the existence of the uncanny in human-like virtual characters, there was still tremendous potential for further research to investigate how factors such as dynamic facial expression, contingency during interaction, timing and sound contribute to perception of uncanniness. Building on this study, and as discussed in the following chapters, my later work investigates much more closely how individual factors such as facial expression and speech may contribute to a more multidimensional model to measure the uncanny.

2.4 LOST IN TRANSLATION?

Generally, there appears to be a wide understanding of the words that Mori (1970/2012) used in his original title for the Uncanny Valley (*bukimi no tani*). In Japanese, *tani* stands for "valley" and *bukimi* stands for "weird, ominous or eerie." In English, words synonymous with the word *uncanny* include "unfamiliar, eerie, strange, bizarre, abnormal, alien, creepy, spine tingling, inducing goose bumps, freakish, ghastly and horrible" (MacDorman and Ishiguro, 2006, p. 312). Hence, the English translation of *Uncanny Valley* was made and accepted without deliberation. However, this generic understanding (and acceptance) of the words that Mori used in his title for the Uncanny Valley does not always apply to those words used as variables to measure the uncanny in synthetic agents. To this day, ambiguity remains as to the actual meaning of Mori's original concept for the Uncanny Valley to the extent that some have argued that it may have been "lost in translation" (Bartneck et al., 2009, p. 270).

One underlying concern raised by many researchers (including myself) investigating the concept of the uncanny in synthetic agents was that of the actual accuracy and reliability of the words used to describe and measure perceived uncanniness. In other words, how do we define our perceived comfort level with a human-like synthetic agent? A word was required that summarized not only how we felt toward an object but, importantly, also what we did *not* feel toward that object. Even Mori himself struggled with this dilemma, and as a solution he created a new word in the Japanese language, namely, *Shinwakan*. However, because it is an uncommon word in Japanese culture, there is no direct translation of this word in English. In an attempt to explain and translate *Shinwakan*, the word "familiarity" is frequently used because it is the opposite of the word "unfamiliarity," which is one of the synonyms for *bukimi*. The word "strange" is a typical word used to describe what is unfamiliar, and many researchers (including myself) have used it as a more negative description of perceived uncanniness to counterbalance the more positive description of "perceived familiarity" when measuring the uncanny (see, e.g., Bartneck et al., 2009; Hanson, 2006; MacDorman, 2006; Seyama and Nagayama, 2007; Tinwell, 2009; Tinwell, Grimshaw, and Williams, 2011). Despite this, many have questioned the actual meaning of familiarity as it lends itself open to misinterpretation. For example, it may be construed as to how well-known one perceives a character to be in an animation or game or, in the case of an android replica, how famous that person is perceived to be. Furthermore, one may interpret the word familiarity to rate how familiar they are with a certain type of character, such as anthropomorphic-type characters, human-like virtual characters or zombies from their previous interaction with such characters in animation or games.

As shown in the previous section of this chapter, I have also used the word familiarity to measure the uncanny in my work on the Uncanny Valley mainly for two reasons: first, because this was the common translation of *Shinwakan* that Mori originally used to describe the uncanny; and second, because other researchers who inspired my own work (such as Karl F. MacDorman, David Hanson and Jun'ichiro Seyama) also used the term in experiments as a dependent variable to measure the uncanny. In my initial "Uncanny as Usability Obstacle" study (and my later work), I explained to all participants in experiments that "familiarity should be interpreted to describe objects that did not appear strange or odd, as opposed to whether an object was well-known" (Tinwell et al., 2011, p. 333). However, despite these instructions that characters should be rated as to how authentically believable their appearance and behavior were perceived to be and not how well-known they appeared, one still cannot ignore a potential misinterpretation of familiarity, especially when rating well-known characters such as Sonic, Mario and Lara Croft (see, e.g., Tinwell, 2009). This confounding factor may at best serve as an (acknowledged) limitation in studies but at worst may risk the intrinsic properties of the Uncanny Valley phenomenon being missed.

The appropriateness of the word familiarity is also a pressing issue for those working with androids. Having identified a significant relationship between the words familiarity and likeability, Christopher Bartneck and colleagues (2009) argued that in the context of the uncanny, perceived likeability was much closer to what Mori had originally meant:

> Our translation of *Shinwa-kan* is different from the more popular translation of 'familiarity.' We dare to claim that our translation is closer to Mori's original intention and the results show that the correlation of likeability to familiarity is high enough to allow for comparison with other studies. (p. 275)

The fact that Bartneck chose to use the word *dare* when making his claim helps to demonstrate that the word familiarly, although not perfect, was widely used as a standardized measure for the uncanny in synthetic agents. Thus, challenging the use of familiarity may have risked a lack of standardization (and cohesion) across different studies as researchers may no longer look to measure the uncanny using this popular translation. One other possibility that Bartneck et al. suggested was to use *Shinwakan* as a technical term in its own right, yet, as they acknowledged, interpretation of this word may prove even more complicated for participants (and researchers) than familiarity. Rather than put researchers off investigating the uncanny, it seems that Bartneck et al.'s brave proposition gave other researchers the confidence to explore other words outside familiarity and to devise more accurate scales to measure perceived uncanniness. Building on Bartneck's work, in 2010 Chin-Chang Ho and Karl F. MacDorman chose to revisit Mori's original theory of the Uncanny Valley, and they found that the indices attractiveness, eeriness, humanness and warmth were more reliable indicators than familiarity to capture the essence of the uncanny in experiments (p. 1517). As you will see in my later work, I too become ever more discerning of familiarity as an adequate and/or appropriate description of perceived uncanniness, and, as my confidence grew along with that of others researching this field, I included many more variables in experiments to measure uncanniness. Scales such as perceived cold-heartedness–warm-heartedness, reassuring–eerie, repulsive–agreeable (Ho and MacDorman, 2010, p. 1514) and apathetic–responsive (Bartneck et al. 2009, p. 79) were also included in my later experiments to help better determine personality traits that may underlie perception of the uncanny in a human-like virtual character.

The neologistic approach that Mori took when creating the word *Shinwakan* may have satisfied his original, hypothetical diagram of the

Uncanny Valley. However, as empirical studies have revealed, it failed to cover *all* aspects of perceived uncanniness as quantifiable, viable and measurable dependent variables. Researchers in this field do acknowledge that investigation into the Uncanny Valley may be more robust if a standard word were used for the dependent variable to describe and measure perceived uncanniness. If just one word could be settled upon, then comparison of results across different experiments and using different stimuli (i.e., from android to virtual character) may be more attainable. On the other hand, using just one word may limit or negate important factors that need to be identified to increase our understanding of the Uncanny Valley. This remains a contentious issue for those researching the Uncanny Valley—but a challenge that many are still willing to pursue.

2.5 THE EFFECT OF AGE AND GENDER ON SENSITIVITY TO THE UNCANNY VALLEY

Previous studies have revealed that participant demographics such as age and gender do impact sensitivity to the uncanny in human-like synthetic agents. The robot engineer and researcher Takashi Minato at Osaka University and colleagues (2004) conducted an experiment in which male and female Japanese adults and children interacted with an android named Repliee R1. The adults were undergraduate and postgraduate university students, and the children were of preschool age between three and five years old. Repliee R1 was designed with the appearance and behavior of a five-year-old Japanese girl. When participants were presented with this android as an interlocutor, in other words, to hold a one-on-one conversation with the android, sensitivity to the uncanny was much stronger for the children than for the adults. This was to the extent that some children were scared of the android and were unwilling to interact or engage in conversation with it. The authors suggested that overall children were more sensitive to the uncanny than adults due to the effect of habituation. Adults may have felt more comfortable with the android as they may have previously seen pictures of similar androids or even come face to face with an android before (Minato et al., 2004). Therefore, the adults may have been comparatively less wary or afraid of the android than the children as they had become more familiar and accepting of androids over time and could better rationalize the android's strange and unnatural behavior. I suggest that by simulating the appearance and behavior of a young girl, Repliee 1 may have presented less of a perceived threat to the adults than to the children who were of a similar age to the android. Given the increase of characters with near human-like appearances being used in children's animation and games, empirical and theoretical work still needs to be done to compare the response of children to that of adults to uncanny, human-like virtual characters.

As Chin-Chang Ho and Karl F. MacDorman (2010) identified, interpersonal warmth lies at the essence of describing (and measuring) the emotional response of the viewer toward synthetic agents. In everyday social behavior, perceived warmth accounts for over half of perceived social traits in others (and oneself) and is a dominant factor of human social perception (Fiske, Cuddy and Glick, 2007; Ho and MacDorman, 2010). This aspect of measuring the uncanny may help us better understand and explain gender difference and sensitivity to the uncanny in adults and children. Earlier work to investigate viewer perception of the uncanny in robots and virtual characters has revealed that males may be more sensitive to the uncanny in near human-like synthetic agents, while females are more accepting of such characters (see, e.g., Green et al., 2008; MacDorman et al., 2010; Tung, 2011). The results of the study undertaken by Green and colleagues (2008) that assessed facial proportion and the uncanny in robots, humans and human-like virtual characters showed that, overall, women were more accepting and tolerant of faces that deviated from the best facial proportions than men. Women were also more likely to regard the robot stimuli as more human-like than men (Green et al., 2008). The authors suggested that this finding may have been explained by "traditional female nurturing roles" (p. 2472) causing females to be more accepting of nonhuman-like traits in robots and more permissive of flaws presented in character's facial features. In this way, women may be more likely to anthropomorphize nonhuman-like robots and project human-like feelings and thoughts onto the synthetic objects. Conversely, men may be more likely to regard the robots as mechanical tools without the ability to think and feel as a fellow human would. These results and theories are supported by the findings from a study in which female human-like, virtual characters were used as stimuli that was orchestrated by MacDorman et al. (2010). They wished to investigate how presentational factors such as motion quality may affect uncanniness in human-like virtual characters. To do this, videos of a woman narrating an ethical dilemma and a computer-generated (CG) 3D replication of this woman's movements and speech were created: then, an additional set of videos was made, but this latter set was manipulated by removing frames in the video. Therefore, as well as being presented with the original videos showing characters with smooth body and facial movements, participants were also presented with videos of the same stimuli, but this time with jerky motion. Interestingly, the stimulus condition (i.e., whether the character was human or not or had smooth or jerky motion) made little difference to women participants. However, males were significantly more likely to bias *against* the female human-like virtual character and especially so when presented with jerky motion. Similar to Green et al.'s (2008) explanation for this gender difference, MacDorman et al. (2010) suggested that men may be less responsive than women to human-like traits in virtual characters and also less forgiving of deviations from the human norm, despite increased levels in realism in the virtual characters.

This effect of gender on perception of uncanniness in adults also occurs in children. A study by Fang-Wu Tung (2011) showed that girls aged eight to fourteen years were more accepting of android robots than boys, especially female androids. Tung (2011) suggested that this may have occurred as girls were more open to accepting robots as social companions than boys. Despite a reduced human-like appearance from the human norm, girls were better able to attribute human-like traits to that robot. This ability to anthropomorphize the androids helped the girls better perceive the androids as social partners than boys (Tung, 2011).

However, the question remains as to why overall, females are better able to assert human-like traits onto human-like synthetic agents? In all of these studies that have explored the effect of participant gender and the uncanny, it seems that a common explanation is offered as to why females may be less sensitive to the uncanny than males. The authors suggested that this may be due to stereotypical traditional female nurturing roles in society that instigate a more forbearing nature in females and hence, a more accepting attitude toward human-like synthetic agents (Green et al., 2008; MacDorman et al., 2010; Tung, 2011). This possible explanation is supported in literature on traditional gender role stereotypes in adults and children. From an early age, girls are commonly taught to be more aware of others' needs than boys (Briton and Hall, 1995; Noller, 1986). To help girls achieve this awareness of others, they may become more perceptive of and responsive to another's emotional state compared to boys. While girls are typically encouraged to focus on more caregiving and social relationships, boys may be more likely to be encouraged by their family, friends and teachers to concentrate on more goal-orientated and competitive relationships (Gilligan, 1982; Perren et al., 2007). As they get older, women may further develop these traits as a survival function for themselves and others in society. For example, a mother may be able to use these nurturing and caregiving skills to encourage a family to stay together or be less likely to reject a child born with defects from the human norm. As such, these characteristics may make women and girls typically more accepting of human-like synthetic characters than males who may be less tolerant of imperfections in these characters' appearance and behavior (Green et al., 2008; MacDorman et al., 2010; Tung, 2011).

2.6 AN EVOLUTIONARY OR DEVELOPMENTAL PHENOMENON?

A question that some researchers have considered is whether experience of the uncanny in human-like synthetic agents is developmental or evolutionary in origin. In other words, do we experience uncanniness due to nature or nurture? In 2012, David Lewkowicz and Asif Ghazanfar wished to challenge the existing assumption that the uncanny is evolutionary and that we are born with the skills to detect when something is not quite right in a human (or object that is attempting to mimic a human). Instead they proposed that we develop this skill to detect abnormalities in another's behavior from infancy and that early experience with human faces is crucial to develop sensitivity to the uncanny. To test this, they conducted an experiment in which groups of 6-, 8-, 10- and 12-month-old infants viewed images of the faces of (1) humans, (2) realistic, human-like virtual characters and (3) uncanny human-like virtual characters with an abnormal appearance. Building on the previous research of Jun'ichiro Seyama and Ruth Nagayama (2007), the eye size of the uncanny virtual characters was increased to a disproportionate, abnormal size so that the characters would appear strange and uncanny. Lewkowicz and Ghazanfar argued that, given that experience of the uncanny relies upon a viewer's ability to detect (relatively) subtle imperfections in a human-like synthetic agent and that perceptual expertise of another's appearance and behavior emerges in the latter half of an infant's first year of life, it is during this time that humans develop an awareness of the uncanny in others who deviate from the human norm. The results of their experiment supported their argument that infants of more than six months old spent longer looking at the human face than the realistic, human-like virtual character and the uncanny character. Therefore, their visual preference was for the human face and not the synthetic, or uncanny, human-like characters. However, infants of up to six months of age spent longest looking at the uncanny face over the human face. Collectively, these findings suggested that sensitivity to the uncanny may be absent in children of up to six months of age and gradually emerges in the second half of an infant's first year of life, at the same time as their perceptual expertise is being developed.

No less than 12 months later, an experiment by Yoshi-Taka Matsuda and colleagues (2012) based on the concept of perceptual narrowing in infants provided support for Lewkowicz and Ghazanfar's (2012) theory, in that instead of the uncanny being an innate instinctive reaction, it is developmentally determined, over time. Perceptual narrowing is a process that develops from birth, as an infant gradually learns to discriminate against different face types (e.g., human or monkey) and to recognize the faces most familiar to them such as their main caregiver (Pascalis et al., 2002). Matsuda et al. found that when infants between 7 and 12 months of age were shown an image of their mother's face or that of a stranger, alongside a CG image that morphed half of their mother's face with half of the stranger's face, the infants avoided looking at the morphed face and preferred the image of the mother or stranger. Importantly, this effect was stronger in those infants of increasing age of up to 12 months. These results showed that infants may detect and avoid faces that appear abnormal (in this case, the morphed image) out of preference to either a familiar or novel face. Matsuda suggested that this uncanny response in the infants was due to the processes of perceptual learning and the ability to differentiate between faces as well as the failure of the morphed image to meet the normal expectations associated with the mother's face (Matsuda et al., 2012).

I put forward that the concept of perceptual narrowing that we have developed with experience of other people's faces from birth may help toward explaining audience response to the uncanny that is regularly reported in films using CG simulations of well-known actors or actresses. For example, as discussed in the previous chapter of this book, CG representations of actors Angelina Jolie and Anthony Hopkins were featured in the film *Beowulf* (Zemeckis, 2007). However, the audience failed to be impressed by these synthetic representations, and the distinct qualities and idiosyncrasies of these celebrities that the audience had grown familiar with and could recognize and relate to were simply lost in the CG characters. The actors appeared onscreen but lacked the distinguishing traits that their roles demanded. Similarly, this effect may occur with video games that feature human-like avatars of famous people, such as international footballers (soccer players) featured in the *FIFA* series (EA Sports, 1993–2013): Electronic Arts promises a real-world football experience, yet some players may be put off as they are alerted to the unusual and *uncharacteristic* appearance and behavior of the CG simulations of their football idols. Based on our ability for perceptual expertise and perceptual narrowing, rather than a CG representation of a famous celebrity, or someone familiar to us being used in an animation or game, we may prefer original video footage of either that celebrity or a new person who is less familiar to us. CG characters created using mo-cap technologies go against our perceptual narrowing skills, in that they may alert us to a perceived difference about that character from what we had already learned or memorized about them. Furthermore, our perceptual narrowing skills improve with age, so we effectively become face experts (Nelson, 2001). Therefore, as we progress to adulthood, we may become ever more discerning (and aware) of uncanny human-like virtual characters, not less. As I reveal in later chapters of this book, my explanations for experience of the uncanny build on not only our early perceptual expertise but also other facets that we develop in infancy, such as avoidance strategies of those who may present a threat, facial mimicry and attachment theory in humans. Collectively, this previous work by myself and others supported the claim made by others that Mori's notion may be too simplistic with various factors (including dynamic facial expression and speech) influencing how uncanny a character is perceived to be. While the previous work discussed in this chapter has provided some insight as to building a conceptual framework of the Uncanny Valley in human-like virtual characters, work still needed to be done to investigate much more closely how aspects of a character's facial expression and speech may influence perception of the Uncanny Valley. In the next chapter, I discuss the findings from an experiment that I designed to investigate how cross-modal communication may influence perception of the uncanny in empathetic and antipathetic characters. A set of heuristic guidelines was also established as to how a designer may strategically use factors such as facial expression and speech to design for or against the uncanny.

REFERENCES

Alone in the Dark (2009) [Computer game], Atari Interactive Inc. (Developer), Sunnyvale, CA: Atari Group.

Bartneck, C., Kanda, T., Ishiguro, H. and Hagita, N. (2009) "My robotic doppelganger—A critical look at the Uncanny Valley theory," in *Proceedings of the 18th IEEE International Symposium on Robot and Human Interactive Communication*, Japan, pp. 269–276.

Blizzard Entertainment (2004). *World of Warcraft* [PC/MMOG], Irvine, CA: Blizzard Entertainment.

Briton, N. J. and Hall, J. A. (1995) "Beliefs about female and male nonverbal communication," *Sex Roles*, vol. 32, pp. 79–90.

Duffy, B. (2003) "Anthropomorphism and the social robot," *Robotics and Autonomous Systems*, vol. 42, no. 3–4, pp. 177–190.

FIFA Football (1993–2013) [Computer game], EA Sports (Developer), Redwood City, CA: Electronic Arts.

Fiske, S. T., Cuddy, A. J. C. and Glick, P. (2007) "Universal dimensions of social cognition: Warmth and competence," *Trends in Cognitive Sciences*, vol. 11, no. 2, pp. 77–83.

Gilligan, C. (1982) *In a Different Voice*. Cambridge, MA: Harvard University Press.

Grand Theft Auto IV (2008) [Computer game], Rockstar North (Developer), New York: Rockstar Games.

Green, R. D., MacDorman, K. F., Ho, C.-C. and Vasudevan, S. K. (2008) "Sensitivity to the proportions of faces that vary in human likeness," *Computers in Human Behavior*, vol. 24, no. 5, pp. 2456–2474.

Hanson, D. (2006) "Exploring the aesthetic range for humanoid robots," in *Proceedings of the ICCS/CogSci-2006 Long Symposium: Toward Social Mechanisms of Android Science*, Vancouver, Canada, pp. 16–20.

Ho, C.-C. and MacDorman, K. F. (2010) "Revisiting the uncanny valley theory: Developing and validating an alternative to the Godspeed indices," *Computers in Human Behavior*, vol. 26, no. 6, pp. 1508–1518.

Ho, C.-C., MacDorman, K. F. and Pramono, Z. A. (2008) "Human emotion and the uncanny valley. A GLM, MDS, and ISOMAP analysis of robot video ratings," in *Proceedings of the Third ACM/IEEE International Conference on Human-Robot Interaction*, Amsterdam, pp. 169–176.

Hodgins, J., Jörg, S., O'Sullivan, C., Park, S. I. and Mahler, M. (2010) "The saliency of anomalies in animated human characters," *ACM Transactions on Applied Perception*, vol. 7, no. 4, Article No. 22.

Image Metrics (2008) *Emily Project* [Facial animation]. Santa Monica, CA: Image Metrics, Ltd.

Image Metrics (2008) *Warrior* [Facial animation]. Santa Monica, CA: Image Metrics, Ltd.

Kageki, N. (2012) "An uncanny mind: Masahiro Mori on the Uncanny Valley and beyond," *IEEE Spectrum*. Retrieved December 2, 2013, from http://spectrum. ieee.org/automaton/robotics/humanoids/an-uncanny-mind-masahiro-mori-on-the-uncanny-valley.

Kanda, T., Hirano, T., Eaton, D. and Ishiguro, H. (2004) "Interactive robots as social partners and peer tutors for children: A field trial," *Human Computer Interaction* (Special issues on human–robot interaction), vol. 19, pp. 61–84.

Lara Croft Tomb Raider: Legend (2006) [Computer game], Crystal Dynamics (Developer), London: Eidos Interactive.

Lewkowicz, D. J. and Ghazanfar, A. A. (2012) "The development of the uncanny valley in infants," *Developmental Psychobiology*, vol. 54, pp. 124–132.

LittleBigPlanet (2008) [Computer game], Media Molecule (Developer), London: Sony Computer Entertainment.

MacDorman, K. F. (2006) "Subjective ratings of robot video clips for human likeness, familiarity, and eeriness: An exploration of the Uncanny Valley," in *Proceedings of ICCS/CogSci-2006 Long Symposium: Toward Social Mechanisms of Android Science.*, Vancouver, pp. 26–29.

MacDorman, K. F. and Ishiguro, H. (2006) "The uncanny advantage of using androids in cognitive and social science research," *Interaction Studies*, vol. 7, no. 3, pp. 297–337.

MacDorman, K. F., Coram, J. A., Ho, C.-C. and Patel, H. (2010) "Gender differences in the impact of presentational factors in human character animation on decisions in ethical dilemmas," *Presence: Teleoperators and Virtual Environments*, vol. 19, no. 3, pp. 213–229.

Matsuda, Y.-T., Okamoto, Y., Ida, M., Okanoya, K. and Myowa-Yamakoshi, M. (2012) "Infants prefer the faces of strangers or mothers to morphed faces: An uncanny valley between social novelty and familiarity," *Biol Lett*, vol. 8, pp. 725–728.

Mega Man (1987–2013) [Computer game], Capcom (Developer), Japan: Capcom.

Minato, T., Shimada, M., Ishiguro, H. and Itakura, S. (2004) "Development of an android robot for studying human-robot interaction, in Orchard, R., Yang, C. and Moonis, A. (eds.), *Proceedings of the 17th International Conference on Industrial and Engineering Applications of Artificial Intelligence and Expert Systems.* Heidelberg: Springer, pp. 424–434.

Mori, M. (2012) "The uncanny valley" (MacDorman K. F. and Kageki, N., trans.), *IEEE Robotics and Automation*, vol. 19, no. 2, pp. 98–100. (Original work published in 1970.)

Nelson, C. A. (2001) "The development and neural basis of face recognition," *Infant and Child Development*, vol. 10, pp. 3–18.

Noller, P. (1986) "Sex differences in nonverbal communication: Advantage lost or supremacy regained?" *Australian Journal of Psychology*, vol. 38, pp. 23–32.

Pascalis, O., de Haan, M. and Nelson, C. A. (2002) "Is face processing species-specific during the first year of life?" *Science*, vol. 296, pp. 1321–1323.

Perren, S., Stadelmann, S., von Wyl, A. and von Klitzing, K. (2007) "Pathways of behavioural and emotional symptoms in kindergarten children: What is the role of pro-social behaviour?" *European Child & Adolescence Psychiatry*, vol. 16, pp. 209–214.

Pollick, F. E. (2010) "In search of the uncanny valley," *Lecture Notes of the Institute for Computer Sciences. Social Informatics and Telecommunications Engineering*, vol. 40, no. 4, pp. 69–78.

Richards, J. (2008) "Lifelike animation heralds new era for computer games," *The Times Online*. Retrieved November 29, 2013, from http://www.thetimes.co.uk/tto/technology/article1859020.ece.

Schneider, E., Wang, Y. and Yang, S. (2007) "Exploring the Uncanny Valley with Japanese video game characters," in *Proceedings of Situated Play, DiGRA 2007 Conference*, Tokyo, Japan, pp. 546–549.

Seyama, J. and Nagayama, R. S. (2007) "The uncanny valley: The effect of realism on the impression of artificial human faces," *Presence*, vol. 16, no. 4, pp. 337–351.

Silent Hill Homecoming (2003) [Computer game], Double Helix Games (Developer), Japan: Konami.

Stanton, A. (producer/director). (2003). *Finding Nemo* [Motion picture]. Los Angeles, CA: Disney.

Steckenfinger, A. and Ghazanfar, A. (2009) "Monkey behavior falls into the uncanny valley," *Proceedings of the National Academy of Sciences of the United States of America*, vol. 106, no. 43, pp. 18362–18366.

Super Mario Sunshine (2002) [Computer game], Nintendo EAD (Developer), Japan: Nintendo.

The Casting (2006) [A technological demonstration for the computer game, *Heavy Rain*], Quantic Dream.

Tinwell, A. (2009) "The uncanny as usability obstacle," in *Proceedings of the Online Communities and Social Computing Workshop, HCI International 2009*, San Diego, CA, pp. 622–631.

Tinwell, A. and Gimshaw, M. (2009) "Bridging the uncanny: An impossible traverse?" in *Proceedings of the 13th International MindTrek Conference: Everyday Life in the Ubiquitous Era*, Tampere, Finland, pp. 66–73.

Tinwell, A., Grimshaw, M. and Williams, A. (2011) "The Uncanny Wall," *International Journal of Arts and Technology*, vol. 4, no. 3, pp. 326–341.

Tung, F. W. (2011) "Influence of gender and age on the attitudes of children towards humanoid robots," in *Proceedings of the 14th International Conference, HCI International 2011*, Orlando, pp. 637–646.

Vinayagamoorthy, V., Steed, A. and Slater, M. (2004) "An eye gaze model for dyadic interaction in an immersive virtual environment: Practice and experience," *Computer Graphics Forum March 2004*, vol. 23, no. 1, pp. 1–11.

Vinayagamoorthy, V., Steed, A. and Slater, M. (2005) "Building characters: Lessons drawn from virtual environments," in *Proceedings of Toward Social Mechanisms of Android Science, COGSCI 2005*, Stresa, Italy, pp. 119–126.

Weschler, L. (2002) "Why is this man smiling?" *Wired*. Retrieved December 2, 2013, from http://www.wired.com/wired/archive/10.06/face.html.

White, G., McKay, L. and Pollick, F. (2007) "Motion and the uncanny valley," *Journal of Vision*, vol. 7, no. 9, p. 477.

Zemeckis, R. (2007). *Beowulf* [Motion picture]. Los Angeles, CA: ImageMovers.

CHAPTER 3

Survival Horror Characters and the Uncanny

The uncanny is ghostly. It is concerned with the strange, weird and mysterious with a flickering sense (but not conviction) of something supernatural. The uncanny involves feelings of uncertainty, in particular regarding the reality of who one is and what is being experienced.

ROYLE (2011, p. 1)

U P TO THIS POINT, exploration of the Uncanny Valley in human-like virtual characters had produced limited explanation as to how aspects such as facial expression and speech may influence perception of the uncanny. Furthermore, while guidelines had been created that focused on how to reduce the uncanniness in empathetic human-like characters, no one had yet considered how the uncanny may be manipulated to *enhance* the fear factor for antagonist characters, intended to scare and disturb the audience. Antipathetic characters featured in the horror genre in games and films* are designed with the intention to deliberately gratify the pleasure we can seek in frightening ourselves. In the survival horror genre, freakish, monstrous characters may seek out, confront and destroy

---

* In this chapter the word *film* will refer to both film and animation.

players, who rely upon weaponry, skill and survival tactics to escape from or defend themselves against these monster attacks. Set in Pennsylvania, the first-person shooter (FPS) *Left 4 Dead* (commonly referred to as "L4D"; Valve, 2008) is centered on a group of civilian survivors who had escaped a pandemic virus that had caused hideous physiological and psychological mutation to the majority of citizens in the city of Fairfield. The survivors included empathetic characters with a human-like appearance such as Louis, an assistant store manager (voiced by Earl Alexander), and Francis, a biker (voiced by Vince Valenzuela). The development team at Valve intended that the survivors would endure endless pursuit by those infected by the virus, represented as horrific, mindless zombie characters with a hideous nonhuman-like appearance. If it was not the gurgling screams of the hoards of the infected, then the formidable, shocking screams of the scrawny, gray emaciated witch or the thunderous pounding and ferocious growls of the hulk-like Tank character signaled an imminent attack.

Developers at Double Helix Games used the technique of anthropomorphism to create hideous antipathetic characters featured in *Silent Hill Homecoming* (2008). The plot follows the plight of the soldier, Alex Shepherd, as he returns to his hometown of Shepherd's Glen to search for his missing father and brother. As an empathetic character, Alex has a human-like appearance, yet various ghastly humanoid mutants with a nonhuman-like appearance haunt this protagonist to try to prevent him from fulfilling his task. During the game, players are confronted with the half-human, half-blade monster the Schism. On top of its disjointed human torso, arms and legs, this character's most distinctive feature is its elongated, pendulum-shaped head that ends in a deadly curved metal blade. One of the most disturbing monsters is a spider-like creature named Scarlet who, with a woman's head and torso, prowls around on arachnid arms and legs with exposed bloody muscle and tissue. Scarlet's wide mouth, filled with needle-like teeth, allows her an increased bite radius for her victims. This character's narration of nonhuman-like snarls, shrieks and snake-like hisses match her arachnid, nonhuman-like appearance.

Yet were designers aware of the strategies that could be taken to make their characters even spookier, more eerie and more spine-tingling? In this chapter I describe an empirical study that I undertook in 2010, now commonly referred to as the Uncanny Modality study, to examine how cross-modal factors such as facial expression and speech may be used to exaggerate uncanniness (see Tinwell, Grimshaw and Williams, 2010). Given that the uncanny is as synonymous with the horror genre as horror with the uncanny, many have identified how the two entities may be strategically used to enhance and complement one another in horror films. In his book *Uncanny Bodies: The Coming of Sound Film and the Origins of the Horror Genre* (2007), the cinema theorist Robert Spadoni attributed the international success of prominent films in the classic horror cycle to the sound cinema timeline. Technical limitations in sound recording and production produced sound that may have been perceived as unnatural and strange. These bizarre sounds increased the banality of these films as the actors and audience adapted to this new medium. Building on Spadoni's observations in early sound cinema, I consider how aspects such as speech quality, lip-synchronization and articulation of speech may exaggerate the uncanny and the ambience of fear in the survival horror genre (Tinwell et al., 2010). The interactive nature of survival horror games and the implications of player interactivity for game designers compared with watching a film are also discussed. To summarize the findings from this inquiry, I provide a set of heuristic guidelines that may be used to control the uncanny so that this effect may work to the advantage of antipathetic zombies and create shock and fear as the designers intended.

3.1 EARLY SOUND CINEMA

The subject of the uncanny is inextricably linked with the horror genre with regards to its cause, function and effects. An object that challenges or contests one's sense of reality and appears strange may alert one to a potential danger or threat, thus evoking an uncomfortable, unsettling feeling to the extent of morbid fear. Tod Browning's *Dracula* (Browning, 1931) and *Frankenstein* (directed by James Whale, 1931) are arguably two of the most famous and iconic monsters in horror cinema, with many film critics proclaiming these characters as timeless. The charismatic Hungarian actor Bela Lugosi was cast as *Dracula*, and, as an auteur, Whale carefully selected the English-born actor Boris Karloff to play Mary Shelley's monster. While many factors may have contributed to the international success of *Dracula* and *Frankenstein* such as the plot, prosthetics, special effects, makeup and actors' performance, Spadoni (2007) attributed it simply to the release dates of these films in relation to the sound cinema timeline. Spadoni argued that an "uncanny body modality of early sound films" (p. 11) occurred between the visual and auditory modalities in the early stages of sound cinema, and it was the particular characteristics of uncanny sound that contributed to the acclaim of these two films. In the transition between silent and sound cinema, technical limitations created speech sound of an unnatural and low quality with blatant discrepancies in the synchronization of speech with lip movement. However, Hollywood film directors in the horror genre were able to exploit these strange and peculiar characteristics of speech to work to the advantage of monsters such as *Frankenstein* and *Dracula*. In this way, a lack of human-like facial expression not only created an ambience of fear and eeriness for *Frankenstein* but also an array of nonhuman-like (almost unidentifiable) grunts and snarls. As well as the gruesome, unnatural blood-curdling textures of Dracula's speech, asynchrony errors increased uncanniness. A distinct lack of synchronization of Dracula's voice with his lip movement represented a disembodiment of his voice with his physical body, thus enhancing the supernatural or ghostly qualities for this monster. Spadoni argued that the unique qualities and timeless appeal of these film characters could not have been achieved without the initial technical flaws and audience conceptual interpretation of this novel medium in early sound cinema. The introduction of sound encouraged strange, ghostly sounds that were beneficial to the eerie and ghostly experience that directors of horror cinema wished to emulate (Spadoni, 2007). Furthermore, Spadoni stressed that these strange benign sounds reminded viewers that the film that they were watching was an artificial, man-made entity. "The introduction of sound resensitized viewers to the artificial nature of cinema, and the resulting resurrection of the forgotten phantoms of the earlier time haunted the sound film screen long enough to shape the beginnings of a Hollywood genre" (p. 121).

Another aspect that Spadoni (2007) observed in his thesis on the uncanny and early sound cinema was the difficulty that actors had adjusting to pronouncing words in a "normal" manner. Without sound, they were instructed to fully emphasize every word and syllable to help the audience understand what was being said. In this way, the audience could lip read mute characters in addition to reading subtitles. It seemed that this technique was a hard habit to break when sound was introduced, and many actors continued to overpronounce words and exaggerate mouth movements to the extent that they appeared absurd or comical. Furthermore, the actors were also encouraged to speak more slowly than usual to aid the sound recording process and to create speech that was clear and intelligible. However, this often led to a rather impeded, delayed delivery style of speech that the audience could perceive as unnatural and odd. Rather than serving as a limitation for Tod Browning's film, Lugosi strategically manipulated this strange, awkward speech delivery style to increase how strange Dracula was perceived to be. The eerie textures of Lugosi's voice were exploited to create a greater conceptual anomaly for the audience. Film critics described the distinctive acoustics and slow and painstaking delivery style of Dracula's speech (with the technique of pronouncing each syllable) as similar to that used by radio broadcasters of the time (p. 64). Spadoni claimed that these resonating speech qualities that disturbed and haunted the audience epitomized the essence of what the "voice of horror" should be (pp. 63–70).

3.2 SURVIVAL HORROR VERSUS HORROR FILM

Where games differ from animation is through their interactive and immersive environments. Video games are an interactive medium that relies upon player input to precipitate events in that game. Environments, characters and sounds may be triggered only as a result of players' actions, which in turn may initiate the inclusion of other environments, characters or sounds. Rather than the more passive interactivity of animation and film where a viewer typically sits back and watches the animation on a screen, a video game player is required to interact with the events on screen via a physical device, such as a joystick or keyboard. Instead of relaxing back in the chair, the player may be on the edge of his seat or may even stand up to establish the most effective and efficient way to interact with the game. This heightened interactivity is often described as player engagement or player immersion, important factors in what constitutes a successful and appealing game (see, e.g., Jennet et al., 2008).

As well as levels of interactivity, other ways films and video games differ is the actual quality of image and sound achievable in real-time games footage compared with that of film's higher resolution and the production process in generating graphics and sound for prerecorded versus real-time footage. The increased number of pixels in film allows for greater graphical detail, and in this way film graphics are always one step ahead of video game graphics and sound files are no different in this respect. Sound files such as a character's speech or sound effect may be optimized for real-time game footage due to limitations in rendering capacity. Reducing the size of the sound files may improve efficiency and download times but impairs the original quality of that sound. Furthermore, the differing production processes involved in sound production for games and film may negate how appropriate sounds are when played at particular points in real-time gaming. In film, speech files are most often recorded as the actor performs, to help ensure an appropriate match between the words spoken and the qualities of speech for a particular scene. However, a video game developer may be limited to a set of prerecorded narratives to utilize within the game despite a difference in context and mood for particular characters during game play. This factor has implications for the authenticity of a particular character if the emotive values of their speech do not match their facial expression or their in-game state. While differences occur in interactivity, quality and production between these two mediums, it is the dire anticipation and adversarial encounters with these monsters that elicits the ultimate, desired response: fear. Building on Spadoni's (2007) theories of uncanny sound in helping to establish Frankenstein and Dracula as horror icons, I realized that no one had considered how the current technical limitations in mo-cap, sound and rendering capabilities may be used to enhance the fear factor in survival horror game characters. The thought came to mind that, whereas the current technological restrictions can exaggerate the uncanny in characters with a highly human-like appearance, why could this not work to the advantage of antipathetic, zombie-type characters in the survival horror game genre as it had previously done for monsters in early sound cinema? Based on these revelations and my own desire to discover more while collecting empirical evidence, I designed an experiment to establish how facial expression, characteristics of speech, lip-synchronization and articulation may be used to control and manipulate the uncanny in character design. A total of 13 videos consisting of antipathetic and empathetic virtual characters with speech narration from film and video games were included in the study, in addition to autonomous social agents and a filmed human.* The set of empathetic human-like characters included Louis and Francis from *L4D*, Alex Shepherd from *Silent Hill Homecoming*, Emily and the Warrior by the computer vision and facial analysis company Image Metrics (2008), and Quantic Dream's Mary Smith. Given that *L4D* was a novel survival horror game at the time that I conducted my experiment in 2009 (before being published the following year), I included four zombie characters from this game: a Smoker, the Infected, the Tank, and the Witch. Footage of a zombie from *Alone in the Dark* (Atari Interactive, Inc. 2009) was also featured in this study due to his more humanoid appearance than other zombies in *L4D*. In addition to footage of a male human, a stylized human-like chatbot librarian named Lillien was included, developed by Daden (2006). To assess perceived uncanniness, 100 participants rated each character video

---

* Full details of the experiment design and results can be found in Tinwell et al. (2010).

(using 9-point Likert-type scales) for how human-like or nonhuman-like and how strange or familiar they judged these characters to be. Specifically, they were asked to rate how human-like the character's facial expression and speech were perceived to be. To assess asynchrony, participants were asked to rate how synchronized the character's voice appeared to be with lip movement, ranging from perfectly synchronized to not synchronized. The participants also were required to select parts of a character's face including the cheeks, forehead, eyes and mouth that appeared to show overexaggerated facial expression or, conversely, a lack of facial expression. To investigate how particular qualities of speech (similar to those renowned by the eponymous Dracula) may effect uncanniness, participants were asked to select if the voice sounded slow, monotone, or of the wrong intonation or if they felt that the voice belonged to that character or not.

The results showed that the five zombie characters and the chatbot, Lillien, were rated close together with low ratings for perceived familiarity and human-likeness (Table 3.1). Ratings for Mary Smith placed her just above the zombie characters but below the mean rating for perceived

TABLE 3.1   Mean Scores For Perceived Familiarity, Human-Likeness, Lip-Synchronization and Human-Likeness for Voice and Facial Expression

Character	Mean Score				
	Familiarity	Human-Like	Lip-Sync	Voice	Expression
The Tank	2.33	2.39	3.08	2.25	2.72
The Witch	2.4	3.28	3.34	2.76	3.04
Lillien (Chatbot)	2.99	3.23	2.51	2.68	2.39
Zombie 1	3	5.24	4.89	4.23	4.88
The Infected	3.1	4.43	3.55	2.88	3.42
The Smoker	3.21	4.19	3.34	2.29	3.32
Mary Smith	4.53	6.63	3.52	7.54	5.37
The Warrior	5.23	7.68	7.35	6.37	7.22
Alex Shepherd	6.38	6.84	5.27	7.78	6.18
Francis	6.8	7.34	6.89	7.75	6.85
Louis	7.02	7.27	7.16	8.04	6.92
Emily	7.25	8.67	8.61	8.8	8.4
Human	8.26	8.8	8.46	8.79	8.6

Source: Tinwell, Grimshaw, and Williams (2010)

Note: Judgments were made on 9-point scales: familiarity (1 = very strange, 9 = very familiar); lip-sync (1 = not synchronized, 9 = perfectly synchronized); human-likeness overall and voice and facial expression (1 = nonhuman-like, 9 = very human-like).

familiarity, while Emily (who had previously been acclaimed as crossing the Uncanny Valley) was still rated below the human for perceived human-likeness and familiarity (Table 3.1).

As discussed in the previous chapter, other authors have suggested that a perceived mismatch in the behavioral fidelity of a synthetic agent with their more human-like appearance evokes the uncanny. The results of this study supported those claims in that human-like characters that were rated as having a lack of human-likeness in their facial expression and speech were regarded less positively and as more uncanny. Mary Smith received a below-average score for how human-like her facial expression was perceived to be, and she may have been rated as less strange and more human-like if the behavioral fidelity for this character matched her more realistic, human-like appearance (Table 3.1). The perceived human-likeness of Emily's facial expression and speech were comparable to that of the human that suggests that this character's behavioral fidelity was more in keeping with the perceived human-like realism of this character (Table 3.1). In the interest of brevity the next sections provide a summary of the main findings from this experiment, but please refer to Tinwell et al. (2010) for a full description of the methodology and results.

3.3 FACIAL EXPRESSION

One of the most iconic features of Frankenstein is the flat metal ridge across his forehead. While the rest of Frankenstein's face is left to move naturally, his upper face is held in place with metal staples, and the rigid, unmalleable qualities of his forehead caused a distinct lack of facial expression for this monster. Spadoni (2007) purported that this attribute raised awareness in the audience of Frankenstein's supernatural otherness. Interestingly, for the virtual characters featured in my experiment, characters were judged as significantly less human-like and familiar when there was a perceived lack of facial expression in the middle and upper regions of the face, such as the eyes and cheeks, but especially so in the forehead. A lack of movement in the forehead was particularly prevalent for the five zombie characters, Lillien and Mary Smith. A viewer may have been reminded of the impulsive, mindless antics of the Tank and the Infected given that there was no facial expression to communicate that character's state of mind. As with Frankenstein, this lack of emotional expressivity may have induced a sense of the diabolical for these characters as viewers may have been confused by their emotional state. Furthermore, they may have been more suspicious of that character as they were not able to decipher how that character was feeling or what it was thinking, hence eliciting the uncanny. This factor, however, also reduced the overall believability and authenticity for the human-like characters Alex Shepherd, Louis and Francis, though less so for Emily and the human. This characteristic may work for zombie characters by enhancing how *abnormal*, strange and uncanny they are perceived to be, but it will work against empathetic characters if a designer has not accurately depicted the upper facial movements of a human-like character correctly.

3.4 SPEECH QUALITIES

The results from the Uncanny Modality study revealed that a slow delivery style of speech significantly increased how nonhuman-like and strange a character was perceived to be. Speech that was judged to be of the wrong pitch and intonation and monotone speech, without intonation and expressivity, were also factors that increased the uncanny for a character. Having identified a strong relationship between participant ratings for "the voice belongs to the character" and "the speech intonation sounds incorrect," I suggested that the overall believability for a character may be reduced if the pitch and intonation of speech are judged to be incorrect based on the character's visual appearance, its actions and, importantly, the context within which that character was placed. This was evidenced by the Tank, who received the lowest mean ratings for perceived familiarity and human-likeness (see Table 3.1), yet the majority of participants judged that his unintelligible snarls and bellowing roars belonged to him. The Tank's nonhuman-like sounds matched his nonhuman-like appearance and were in keeping with his hostile, unfathomable behavior. Similarly, the Witch scored second lowest (see Table 3.1) for perceived familiarity (and only third for human-likeness scores), but her wretched, high-pitched cries and screams were judged to match her scrawny, disfigured, nonhuman appearance. These particular characteristics of speech increased the character's authenticity as there was perceived congruence between their nonhuman-like appearance and their speech fidelity.

However, peculiar speech qualities worked against empathetic characters such as Alex Shepherd, Lillien and Mary Smith, who had originally been designed not to contest a sense of the real and disturb the viewer. Alex Shepherd's voice was perceived as being abnormally slow and of a monotone sound that lacked emotion. These factors did enhance how eerie and frightening the Warrior was perceived to be but unwittingly worked against the protagonist Alex Shepherd: with such retarded and simple intonation, his dull tones did not match his more sophisticated human-like appearance. As found in my 2009 study when virtual character human-likeness ratings were plotted against familiarity ratings (see Chapter 2), the results from the Uncanny Modality study revealed that Lillien rated well below average for perceived familiarity of just 2.99 and was placed second from bottom for human-likeness (see Table 3.1). Given that this result was repeated for Lillien and that she was rated as uncanny as zombie characters, the pressing issue was to establish why this occurred. Daden developed Lillien as a friendly virtual librarian who could answer specific queries about books and library collections. While this information was of help to the interactant, this virtual interlocutor was perceived as more disturbing than helpful. Given that Lillien's voice was one of the most important factors to meet her objectives, her speech received higher than average ratings for perceived slowness, being of an incorrect pitch and monotone. In this way her voice was perceived as dull, monotonous and uninspiring to the point of being a major annoyance to the viewer. Speech that is recorded of a low tempo to aid clarity or produced at a low speed due to automated speech synthesis devices (or both) may be at risk of being perceived as unnatural and strange. Such factors work against how accepting a viewer may be of a character with a human-like appearance, which is intended to be perceived as amiable and friendly. Yet this crippled speech style worked for Zombie 1, who received similar ratings for speech qualities as Lillian. In this case the designers may have intentionally manipulated this zombie's voice to a more laggard, repetitive, unnatural style to increase how eerie and spine-tingling this zombie was portrayed to be.

Mary Smith's speech was judged to be of a more appropriate tempo and less monotone than the aforementioned characters, yet only 20% of participants agreed that her voice actually belonged to her. Participants could perceive emotive tones in her voice; however, there was incongruence between the pitch of her voice with her facial expression and what was happening to her at a particular point in the story. In this case, her vocal tones may have depicted sorrow or fear, yet it was difficult to read such information in her facial expression or body gesture, leaving the viewer confused and unable to empathize with her. Ironically, the increased quality of sound achieved in this tech demo negated Mary's believability to the extent that she was perceived as unnatural and odd. As a possible solution to this problem, I suggested (in a later collaborative work) that it may be best to reduce the level of human-likeness for a character's voice so that the speech fidelity matched the character's behavioral fidelity. However, such modifications would need to be done subtly over a continuum of human-likeness (e.g., from human-like to mechanical) to help increase accord between a character's visual and acoustic modalities.

> Of course we do not suggest that cartoon-like voices be used with characters that are approaching believable realism in computer games, however the level of human-likeness may be subtly modified so that the perceived style of the voice sound matches the aesthetic appearance of the character. (Tinwell, Grimshaw and Williams, 2011, p. 221)

Such tactics may have helped the viewer better accept that the voice actually belonged to Mary Smith rather than detecting a discord between the emotive values of her speech with her more limited appearance and behavior.

3.5 ARTICULATION OF SPEECH

As one of the most complex muscular regions of the human body, the mouth consists of hundreds of individual muscles used to communicate facial expression and speech. As Peter Plantec (2007, p. 4) commented in his appraisal of virtual human-like characters and the Uncanny Valley, the mouth is one of the hardest regions to model in a character, especially during speech: "Often neglected but critically important would be the inside of the mouth. Tongue and teeth/jaw movement is almost as critical as the eyes, and then there's spittle." The reason that the mouth region provides such a complex challenge is that, while we may have attained an understanding of individual muscles in the mouth region, their combined effect is less well understood. Despite high-definition cameras capturing performance data, the mouth is an intricate area to record accurately and may still require modification when being applied to a 3D human-like model. At least with prerecorded footage in film (and game cut scenes and trailers), a 3D modeler may allocate time to edit individual key frames to ensure the most accurate depiction of the actor's mouth movement is achieved. However, this can be a time-consuming process and is an unconventional method for game content generated in real time.

Instead, the designer must rely upon automated processes to configure animated facial expression and lip movement with synthesized speech.

Using the International Phonetic Alphabet as reference, a visual mouth shape (viseme) is created for each phoneme sound that is stored in a video game engine database. An appropriate viseme image is then retrieved from the database to represent different mouth shapes for particular speech sounds such as "p," "th" and "ae." The viseme class titled "normal" is commonly set as default in which character mouth shapes are modeled to represent a normal articulation of speech. Different viseme class types can be created to represent different intensities of speech sound and the associated articulation. For example, the facial animation software *FacePoser* available in the Source Developer Kit (SDK) by Valve (2008) provides the additional viseme classes titled "strong" and "weak" for mouth shapes with an increased or reduced intensity of articulation. The developer may also create new viseme classes bespoke to a particular character and/or emphasis of articulation. Interpolated motion is then generated automatically between each phoneme sound (and viseme image) to achieve multimodal speech synthesis for a character.

Given the complexities of accurately animating the combination of muscles used both inside and outside the mouth during speech, one has to question whether the current technical capacity for synthesized speech processes is sufficient to capture and portray the finer nuances of natural speech. Furthermore, I put forward that incongruence between articulation, tone and volume of speech that was observed in early sound cinema talkies is prevalent to the audiovisual speech synthesis used for real-time speech in video game characters. Actors in early sound cinema were often construed as overacting as they became accustomed to the changes required for mute and sound film. They may still unnecessarily exaggerate their jaw and lip movements as a way to help the viewer understand what was being said, despite the introduction of sound. This overarticulation was regarded as unnatural and comical, and Spadoni (2007) remarked that it needed to be toned down or risk the characters being mocked. Interestingly, the results of the Uncanny Modality study supported Spadoni's observations in early talkies as to the preposterous overexaggeration of the actors' mouth movements. Virtual characters were rated as significantly stranger when they were judged to have an overexaggeration of mouth movements. This effect was significantly more profound for the mouth than in the middle and upper regions of the face including the cheeks, eyes and forehead. Nearly a quarter of participants judged Alex Shepherd to have overexaggerated mouth movements. Given that this character was also rated as having monotone speech, even mouth movements of a normal proportion may have appeared as exaggerated due to the sheer lack of emotive qualities in his speech. In other words, Alex's understated intonation of speech made his mouth movements appear overstated. It may have been beneficial to tone down the intensity of articulation in Alex's mouth movements as the viseme shapes failed to match the intonation of what he said. In this case a standardized speech synthesis process—in other words a one-size-fits-all approach—failed to match appropriate mouth shapes with Alex Shepherd's speech and in game state. Rather than using the normal or strong viseme class, it may have been beneficial to use the weak viseme class or for the designer to create a greater range of individual bespoke mouth shapes for this character, as a better match to his monotone, unemotional speech, especially for game footage generated in real time. The following provides an example of how the facial animation for the word "no" may be construed as out of context:

> The mouth shape for the phoneme used to pronounce the word "no"... may be applicable if the word is pronounced in a strong, authoritative way, but would appear overdone and out of context if the same word was used to provide reassurance in a calming and less domineering manner. (Tinwell et al., 2011, p. 228)

Over a third of participants rated each of the five zombies as having overexaggerated mouth movements, and this factor was of greatest significance for the Tank. This character depicts a blatant lack of control over his jaw, lower lip and tongue movements that appear as banal as his nonhuman-like snarls and roars. One of the reasons that this factor may be so disturbing is that it may infer a lack of motor skills in a character to control mouth movements. Therefore, the Tank's overexaggerated mouth movements may have provided a trigger that the horrific mutations that the virus inflicted had occurred at not only a physical but also a psychological level. In humans, a lack of motor skill control over jaw and lower lip movements is often associated with possible mental dysfunction. For example, a person may have suffered a traumatic head injury or stroke or have developed a degenerative disorder that affects brain nerve cells such as Parkinson's disease. Furthermore, a temporary impairment of motor skills may be caused by excessive alcohol and/or drugs, which suppress the central nervous system. For whatever reason, in the viewer's mind there is the possibility that this character may be in an abnormal mental state, incapable of controlling normal bodily function. Hence, this character now represents a potential threat as the character may be incapable of making reasonable decisions or taking reasonable action. The threat of more uninhibited, impetuous behavior in an antipathetic character may increase how uncomfortable one feels when confronted with such a character. However, abnormal articulatory movement may trigger similar suspicions when presented in empathetic characters (such as Alex Shepherd) to the extent that a viewer may question, "Is this character of sound mind?" which thus works against the character.

3.6 LIP-SYNCHRONIZATION

Spadoni (2007) equated much of the uncanny elicited by the monster Dracula (Browning, 1931) with asynchrony errors between image and sound to the extent that, even though Dracula's lips remained still, evil laughter filled the scene and reverberated through the cinema. Given that the laughter does not appear to project from the mouth of the onscreen character and with no identifiable source of the laughter, it may be perceived as an eerie, disembodied sound. This reminds the viewer of Dracula's ghostly, incorporeal characteristics and the supernatural qualities that he may possess, hence increasing the uncanny for this monster. The results from the Uncanny Modality study revealed that an asynchrony of speech with lip movement was still as important in evoking perception of the uncanny in characters nearly 80 years after *Dracula* and early sound cinema.

The Uncanny Modality study provided the first empirical evidence to show that an asynchrony of speech was directly related to how strange or nonhuman-like a character was perceived to be. Perceived familiarity and human-likeness increased for those characters that were rated as having close to perfect synchronization of sound with lip movement. As ratings for perceived familiarity increased incrementally for Francis, Louis, Emily and the human, so did scores for perceived synchrony of lip movement with speech. Characters rated as least synchronized starting with Lillien, then the Tank, the Witch, the Smoker, Mary Smith and the Infected, respectively, received among the lowest scores for familiarity and human-likeness; thus, asynchrony increased how strange, nonhuman-like and eerie they were perceived to be.

The outcomes from the Uncanny Modality study suggested that the main cause of uncanny speech in video game characters was due to technological limitations in the processes of automated audiovisual speech synthesis production. It appeared that there may not be a sufficient range of viseme (mouth shapes) to completely represent all of the possible combinations of jaw and lip movements available for each phoneme sound at differing levels of intensity. Furthermore, restrictions as to the actual range of facial muscles available within a particular character model may only exacerbate this dilemma further as the designer may fail to match appropriate mouth shapes with particular emotive utterances in differing contexts. This not only applies to the pronunciation of particular phoneme sounds or words at a particular point in a sentence but also to the variation of mouth shapes in the transition between syllables in words or between words when interpolating from one sound to another. This lack of accurate mouth shapes may contribute to a perceived lack of synchrony between lip movement and speech that is commonly reported in uncanny virtual characters.

Much could also be learned from previous standards set in the broadcasting industry as to how asynchrony may affect uncanniness in video games. For television programs, asynchrony errors can occur if there is a delay in transmission times for video and audio signals. Asynchrony errors can cause great annoyance for the viewer and interrupt their enjoyment of a particular program (Reeves and Voelker, 1993). This may cause such confusion and irritation for the viewer that subtitles are regarded more favorably for foreign works instead of using dubbed speech (Hassanpour, 2009). Viewer perception that a sound and image occur at the very same time and from the same source has been defined as *synchresis* by Chion (1994) and *synchrony* by Anderson (1996). However, other researchers have discovered that audiovisual signals do not have to occur simultaneously, at *exactly* the same time, to convince the viewer that they are from the one source (see, e.g., Conrey and Pisoni, 2006; Stein and Meredith, 1993). There is some leverage in our sensitivity to asynchrony to achieve temporal coordination for multisensory signals such as speech and lip movement as a singular event. Previous investigation into synchrony detection in normal-hearing adults has revealed that a synchrony window exists over which asynchronous sound and image can be perceived as synchronous (see, e.g., Conrey and Pisoni, 2006; Dixon and Spitz, 1980; Grant and Greenberg, 2001; Grant, Wassenhove and Poeppel, 2004; Lewkowicz,

1996). Despite experiments having been conducted with participants completing different tasks and using different sound and image stimuli including music, musical instruments and talking heads, two consistent characteristics have been identified for the synchrony window (Conrey and Pisoni, 2006). First, it has a width over several hundred milliseconds wide extending to +400 ms (in which the plus sign denotes that the image precedes sound). Asynchronous material is not immediately identified up to +400 ms, yet is more easily detectable beyond this measurement. Second, it is not symmetrical as viewers are better able to detect synchrony errors when the audio stream precedes the visual stream than when the audio steam lags behind the visual stream. This heightened sensitivity to negative asynchrony (when the audio precedes image) is especially evident in continuous streams of audiovisual speech. An experiment by Grant and colleagues (2004) revealed that participants detected asynchrony when the speech sound was heard at least 50 ms before lip movement was seen. Yet participants were unable to detect asynchrony when the lip movement preceded the speech sound by up to 220 ms. In accordance with these findings, the International telecommunications Union (ITU) recommendations for television broadcast production specify that the video stream should not precede the audio stream by more than 125 ms and should not lag behind the audio stream by more than 45 ms (ITU-R, 1998).

The technique of lip-reading commonly used by those with hearing impediments to aid understanding of what another is saying can help reveal why we can find an asynchrony of speech so confusing and stressful. As well as our interpretation of speech sound, we also rely on the visual information from a person's mouth (Conrey and Pisoni, 2006; Macaluso et al., 2004; Mattys et al., 2000; Murray et al., 2005; Munhall and Vatikiotis-Bateson, 2004). Therefore, a person can listen to what is being said from reading the viseme shapes that a person makes to represent each phoneme sound. However, when the speech sound occurs before we see a person's lips move, our ability to predict what the speaker is going to say using a lip-reading technique may be impeded. Rather than being able to focus on their lips to help us establish what they will say, we are distracted by the speech sound and may wonder where it is coming from. Furthermore, Harry McGurk and John MacDonald (1976) found that a negative asynchrony of speech can actually lead to one misinterpreting what has been said. If a sound syllable is heard first and takes precedence over the visual syllable, this may lead to interpretative conflict so that neither the sound nor image is interpreted correctly. The perceiver may even combine the sound and visual syllable information to create a new syllable. For example, the sound "ba" that coincides with a visual "ga" mouth shape may be interpreted as a "da" sound. This McGurk effect was observed by Karl F. MacDorman in Mary Smith's speech, and he remarked that lip-sync error increased how confusing and disturbing Mary Smith appeared to be. "People 'hear' with their eyes as well as their ears. By this, I mean that if you play an identical sound while looking at a person's lips, the lip movements can cause you to hear the sound differently" (Karl F. MacDorman in an interview with Carrie Gouskos, in *Gamespot*, 2006).

Mary Smith received a significantly lower than average score for perceived synchrony of speech in the Uncanny Modality study, and participants may have experienced the McGurk effect when observing this character, thus exaggerating the uncanny. However, further work was still required to investigate more closely how factors such as the magnitude and direction of asynchrony may affect perception of uncanniness. Therefore, I conducted a further experiment in which 113 participants rated headshot videos of a male human and the character Barney, an empathetic male character with a realistic human-like appearance from the video game *Half-Life 2* (Valve, 2008) over a range of onset audiovisual speech asynchronies for perceived uncanniness.* In keeping with the synchrony window, the videos were manipulated so that AV speech was presented over the following asynchronies: ±400 ms, ±200 ms and 0 ms (the negative value represents asynchronies where the acoustics preceded the video stream). The results revealed that Barney was regarded as more uncanny than the human video and that as the level of asynchrony of speech increased so did perception of uncanniness. Interestingly, participants were significantly more sensitive to the uncanny in Barney when his speech was heard before his lip movement was seen. This result was evident when statistical comparisons were made between negative asynchronous videos and those where Barney's lip movements were seen before his speech was played.

As with humans, when a character's speech occurs before visual lip movement is seen, the viewer's ability to simultaneously process the audio and visual stimuli as one event may be impaired. Under these conditions, the character appears nonsensical as the information normally gained from mouth movement is occluded, therefore possibly preventing accurate

---

* For a full description of the methodology, experiment design and results, please see the original paper by Tinwell, Grimshaw and Abdel Nabi that is currently in press for the *International Journal of the Digital Human*.

comprehension of what the character is saying. To compensate for this intelligible dialogue, the viewer may create a new sound as one struggled to understand what the character was saying. This confusion may abruptly remind a viewer of being presented with a man-made, synthetic object and hence perceived believability for that character is lost. Furthermore, the uncanny may be exaggerated as the viewer perceives that the voice is disembodied from the character, raising suspicion that an unpredictable, ghostly or supernatural force is controlling the actions of that character (Tinwell, Grimshaw and Abdel Nabi, in press).

These findings from my investigations into lip-sync error and perception of the uncanny lend some support to the associations that Jentsch (1906) made of witnessing another person having a seizure with experience of the uncanny. If one perceives that someone may have lost control of their normal bodily or mental functions, such as the seizures induced by epilepsy, this may be frightening for the viewer. The viewer may take a defensive stance against those who present these sudden transitory states of fitting as it may seem that "automated or mechanical processes" were at work behind the individual (p. 226). The uncontrolled babbling effect of asynchronous speech in a character may raise the possibility that the character is no longer in control of their body and mind hence presenting a possible threat to the viewer. This effect may work to the advantage of zombie characters such as the Infected, as an abrupt and shocking realization that the mutations in this zombie have caused mental dysfunction and a lack of bodily control in this monster. This babbling effect, caused by an asynchrony of speech, may be of use for designers to implement when creating characters such as the Infected intended to disturb or frighten the viewer yet may have an adverse effect for viewers attempting to engage with empathetic characters that display such strange behavior. Exploiting the boundaries of the synchrony window in audiovisual speech may work to the advantage of antipathetic characters by enhancing the horror that character may incite. For example, a games engine may be programmed so that antipathetic characters display a negative asynchrony of speech of greater than −400 ms. However, if the uncanny is to be avoided for empathetic characters, then the developers should ensure that, at the very least, any delays in video transmission do not extend beyond that of the sound transmission in game devices. Given the current technical limitations of creating real-time synthesized speech in video games, I put forward that a smaller window of acceptable asynchrony is required for audiovisual speech in virtual characters. With tighter boundaries than that already defined for the television broadcasting industry, this may help to reduce the risk of lip-sync error evoking the uncanny for empathetic game characters (Tinwell et al., in press).

3.7 DESIGNING FOR OR AGAINST THE UNCANNY

Based on the previous empirical and theoretical works discussed in this chapter, I suggest the following guidelines to control the uncanny via facial expression and speech in video game characters. For facial expression:

1. Characters are perceived as more uncanny when there is a lack of human-likeness in the character's facial expression, especially so in the upper facial region, including the forehead and eyebrows. Therefore, a designer should reduce or eradicate movement in the upper facial region to increase the uncanny for antipathetic zombie-type characters using a modeling paradigm based on Frankenstein's monster (Tinwell et al., 2010).

2. An overexaggeration of articulation during speech increased uncanniness in characters. Inappropriate emphasis of mouth movement appeared out of context with the emotive values of a character's speech that may suggest a lack of motor skill control. To reduce the uncanny, articulation may be reduced or carefully modeled so that it matches the emotive qualities of the character's speech. However, to enhance the uncanny and fear factor for a character, a designer should use similar modeling techniques to that of Lugosi's articulation (Tinwell et al., 2010).

The following strategic manipulations may be made to a character's speech:

1. Particular characteristics of speech exaggerated the uncanny in characters, including viewer perception that speech was of an incorrect pitch or tone. Specifically, monotone speech that lacked emotive qualities increased uncanniness to the extent that a voice was no longer perceived as belonging to a character (Tinwell et al., 2010). To reduce the overall believability for a character and to enhance the uncanny, a character's speech may present a lack of emotional expressivity and duller, repetitive acoustics. To avoid the uncanny

and these more disturbing speech qualities, the sound quality and characteristics of speech should match the level of human-likeness for that character (Tinwell et al., 2011).

2. A slow delivery style of speech also evoked the uncanny. Speech played at an unnaturally slow, laggard pace increased how lifeless the character was perceived to be. This tactic may be used to increase the perceived strangeness for zombie-type characters (Tinwell et al., 2010, 2011).

3. Characters were judged as less uncanny when synchrony of lip movement and speech were judged as being close to perfect. However, asynchronous lip-vocalization narration may remind the viewer that the character is simply a manufactured object that cannot be perceived as authentic or believable. Furthermore, a ghostly or supernatural force may be seen to be at work behind the uncanny character. To initiate a presumption that an antipathetic character may be insane or a lunatic, a designer may deliberately increase asynchrony of speech to the extent that the character's speech is unfathomable. Comprehension of speech is impaired further if the speech sound is heard before lip movement is seen, so manipulating lip-sync in this way may increase the absurdity and fear factor for antipathetic characters (Tinwell et al., in press).

While video games champion with increased levels of player interactivity, this factor puts them behind film when comparing the levels of anatomical detail achieved in characters' facial expression and speech. Real-time processing of a character's speech and facial expression can be computationally very demanding and plausibility may be lost in complex game scenarios. As listed already, this can exaggerate the uncanny in many ways. However, some are optimistic that developments in procedural game audio and animation may, at least in part, provide a solution to this pressing problem. Hug (2011) postulates that future procedural sound techniques that rely on haptic feedback will allow for the generation of bespoke sounds for a character in keeping with that character's appearance, behavior and in-game state. Modal synthesis, granulation and Interactive XMF (to name but a few) are dynamic sound generation techniques that will create sounds in real-time responding to aspects such as haptic feedback from the player and the timing and condition of game characters (Hug, 2011). Procedural speech synthesis may allow for a sentence to be generated in game at an appropriate tempo and intonation for a given scenario. The sentence, "I need to find a way out," may be spoken more slowly, in a contemplative manner if a character locked in a chamber is carefully planning an escape route. In contrast, the character may say the same sentence in a rushed, panicked way if being chased by an antagonist character.

Articulation of mouth movement during speech may be improved using procedural animation techniques during speech. Building on the work that has already been done to achieve real-time, data-driven, multimodal procedural generation techniques (e.g., see Cao et al., 2004; Farnell, 2011; Nacke and Grimshaw, 2011), a tool may be developed that combines haptic input from the player, such as their actions or psychophysiology, with the emotive qualities and characteristics of characters' speech and facial expression. For example, the more scared a player is perceived to be due to an increase in GSR readings or heart rate, the more scared an empathetic character's speech and facial expression may be portrayed in-game. Such a tool may help the overall believability for empathetic characters as it is more in keeping with the game context at a given point. Characters such as Alex Shepherd or Mary Smith or chatbots such as Lillian may be perceived as less uncanny if characteristics of their speech and facial expression vary in response to the player's state. Accordingly, a zombie's speech may be played more slowly, with reduced expressivity, exaggerated articulation and lip-sync error if the input from the player is showing increased fear at a particular point to improve the overall player experience in a survival horror game.

In closing this section it is apt to return to the very start: Nicholas Royle's (2011) description of the uncanny (placed at the beginning of this chapter) asserts that the uncanny is ghostly and is related to the supernatural. The first part of Royle's description is particularly poignant to the purpose of this chapter in examining the inextricable link between the uncanny, early sound cinema and the survival horror game genre. Then, as I establish my theories as to why we experience the uncanny and its cause, function and effects, Royle's affiliation of the uncanny as to an uncertainty of who one is becomes ever more prevalent.

REFERENCES

Alone in the Dark (2009) [Computer game], Atari Interactive Inc. (Developer), Sunnyvale, CA: Atari Group.

Anderson, J. D. (1996) *The Reality of Illusion: An Ecological Approach to Cognitive Film Theory*. Carbondale: Southern Illinois University Press.

Browning, T. (1931) *Dracula* [Motion picture]. Universal City, CA: Universal Pictures.

Cao, Y., Faloustsos, P., Kohler, E. and Pighin, F. (2004) "Real-time speech motion synthesis from recorded motions," in Boulic, R. and Pai, D. K. (eds.), *Eurographics/ACM SIGGRAPH Symposium on Computer Animation*, pp. 345–353.

Chion, M. (1994) *Audio-Vision: Sound on Screen* (Gorbman, C., trans.). New York: Columbia University Press.

Conrey, E. and Pisoni, D. (2006) "Auditory-visual speech perception and synchrony detection for speech and nonspeech signals," *Journal of the Acoustical Society of America*, vol. 119, pp. 4065–4073.

Dixon, N. F. and Spitz, L. (1980) "The detection of auditory visual desynchrony," *Perception*, vol. 9, pp. 719–721.

Farnell, A. (2011) "Behaviour, structure and causality in procedural audio," in Grimshaw, M. (ed.), *Game Sound Technology and Player Interaction: Concepts and Developments*, Hershey, PA: IGI Global, pp. 313–339.

Gouskos, C. (2006) "The depths of the Uncanny Valley," *Gamespot*. Retrieved January 5, 2014, from http://uk.gamespot.com/features/6153667/index.html.

Grant, K. W. and Greenberg, S. (2001) "Speech intelligibility derived from asynchronous processing of auditory-visual information," paper presented at the ISCA International Conference on Auditory-Visual Speech Processing, Scheelsminde, Denmark.

Grant, W., Wassenhove, V. and Poeppel, D. (2004) "Detection of auditory (cross-spectral) and auditory–visual (cross-modal) synchrony," *Speech Communication*, vol. 44, pp. 43–53.

Half-Life 2 (2008) [Computer game]. Valve Corporation (Developer), Redwood City, CA: EA Games.

Hassanpour, A. (2009) "Dubbing," *Museum of Broadcast Communications*. Retrieved January 5, 2014, from http://www.museum.tv/eotv/dubbing.htm.

Hug, D. (2011) "New wine in new skins: Sketching the future of game sound design," in Grimshaw, M. (ed.), *Game Sound Technology and Player Interaction: Concepts and Developments*, Hershey, PA: IGI Global, pp. 384–415.

Image Metrics (2008) *Emily Project* [Facial animation]. Santa Monica, CA: Image Metrics, Ltd.

Image Metrics (2008) *Warrior* [Facial animation]. Santa Monica, CA: Image Metrics, Ltd.

ITU Recommendations for Broadcast Television Production (ITU-R) (1998) "BT. 1359: Relative timing of sound and vision for broadcasting (Question ITU-R 35/11)" Retrieved January 5, 2014, from http://www.itu.int/rec/R-REC-BT.1359/en.

Jennett, C., Cox, A. L., Cairns, P., Dhoparee, S., Epps, A., Tijs, T., et al. (2008) "Measuring and defining the experience of immersion in games," *International Journal of Human-Computer Studies*, vol. 66, pp. 641–661.

Jentsch, E. (1906) "On the psychology of the uncanny," *Angelaki* (trans. Sellars, R. 1997), vol. 2, no. 1, pp. 7–16.

Left 4 Dead (2008) [Computer game], Valve (Developer), Washington, DC: Valve Corporation.

Lewkowicz, D. J. (1996) "Perception of auditory-visual temporal synchrony in human infants," *Journal of Experimental Psychology, Human Perception and Performance*, vol. 22, pp. 1094–1106.

Lillian (2006) [A natural language library interface and library 2.0 chatbot]. Birmingham, UK: Daden Limited.

Macaluso, E., George, N., Dolan, R., Spence, C. and Driver, J. (2004) "Spatial and temporal factors during processing of audiovisual speech: A PET study," *NeuroImage,* vol. 21, pp. 725–732.

Mattys, S., Bernstein, L. E., Edward, T. and Auer, J. (2000) "When lipreading words is as accurate as listening," paper presented at the 139th ASA Meeting, Atlanta, GA.

McGurk, H. and MacDonald, J. (1976) "Hearing lips and seeing voices," *Nature*, vol. 264, no. 5568, pp. 746–748.

Munhall, K. and Vatikiotis-Bateson, E. (2004) "Spatial and temporal constraints on audiovisual speech perception," in Calvert, G., Spence, C. and Stein, B. E. (eds.), *The Handbook of Multisensory Processes*, Cambridge, MA: MIT Press, pp. 177–188.

Murray, M. M., Molholm, S., Michel, C. M., Heslenteld, D. J., Ritter, W., Javitt, D. C., et al. (2005) "Grabbing your ear: Rapid auditorysomatosensory multi-sensory interactions in low-level sensory cortices are not constrained by stimulus alignment," *Cerebral Cortex*, vol. 15, pp. 963–974.

Nacke, L. and Grimshaw, M. (2011) "Player-game interaction through affective sound," in Grimshaw, M. (ed.), *Game Sound Technology and Player Interaction: Concepts and Developments*, Hershey, PA: IGI Global, pp. 264–285.

Plantec, P. (2007) "Crossing the great Uncanny Valley." Retrieved January 5, 2014, from http://www.awn.com/articles/production/crossing-great-uncanny-valley.

Reeves, B. and Voelker, D. (1993) "Effects of audiovideo asynchrony on viewer's memory, evaluation of content and detection ability (Research report prepared for Pixel Instruments, CA). Palo Alto, CA: Stanford University, Department of Communication.

Royle, N. (2011) *The Uncanny*. Manchester, UK: Manchester University Press.

Silent Hill Homecoming. (2008) [Computer game], Japan: Double Helix Games (Developer).

Spadoni, R. (2007) *Uncanny Bodies: The Coming of Sound Film and the Origins of the Horror Genre.* Berkeley: University of California Press.

Stein, B. and Meredith, M. A. (1993) *The Merging of the Senses.* Cambridge, MA: MIT Press.

Tinwell, A., Grimshaw, M. and Abdel Nabi, D. (in press) "The effect of onset asynchrony in audio-visual speech and the Uncanny Valley in virtual characters," *International Journal of the Digital Human.*

Tinwell, A., Grimshaw, M. and Williams, A. (2010) "Uncanny behaviour in survival horror games," *Journal of Gaming and Virtual Worlds*, vol. 2, no. 1, pp. 3–25.

Tinwell, A., Grimshaw, M. and Williams, A. (2011) "Uncanny speech," in Grimshaw, M. (ed.), *Game Sound Technology and Player Interaction: Concepts and Developments*, Hershey, PA: IGI Global, pp. 213–234.

Valve (2008) *Faceposer* [Facial animation software as part of source SDK video game engine]. Washington, DC: Valve Corporation.

Whale, James. (1931) *Frankenstein* [Motion picture]. Universal City, CA: Universal Pictures.

CHAPTER 4

Uncanny Facial Expression of Emotion

Words are not emotions, but representations of emotions.

EKMAN (2004, p. 45)

T HE RESULTS OF THE UNCANNY MODALITY STUDY (Tinwell, Grimshaw and Williams, 2010) that I discussed in the previous chapter indicated that a perceived lack of human-likeness in facial expression exaggerated the uncanny in virtual characters with a human-like appearance. Protagonist characters that were intended to be perceived as empathetic, such as Alex Shepherd from *Alone in the Dark* (Atari Interactive, Inc. 2009) and Mary Smith from *The Casting* (Quantic Dream, 2006), were regarded as less familiar and human-like when they were judged to portray a lack of facial emotional expressivity. This effect was particularly salient in the upper facial region, including the eyebrows and forehead. While this finding provided some evidence as to which factors evoked perception of the uncanny, I wanted to know more. It is well established that in humans facial expression is used not only to communicate how one is feeling but also as a way to determine the affective state and possible actions of others (Darwin, 1872; Ekman, 1979, 1992). Importantly, each emotion type serves a different adaptive function as part of social or survival interaction. Therefore, we may respond more negatively to perception of anger or fear in another yet more positively to perceived happiness. Based on this, I pondered whether the uncanny effect would be the same or different across all

71

emotions. If response to the uncanny was to differ across different emotion types, then why would this occur? Also, what would the implications of this be for designers when modeling different emotion types in a character's facial expression? To begin toward working to find an answer to these questions, I set about a new empirical study to investigate more closely how inadequate movement in the upper face may influence perception of a character's emotive state and the uncanny across the six basic emotions (Ekman, 1972). This chapter provides a description of the methodology, design and findings from that experiment to investigate the implications of a lack of nonverbal communication (NVC) in the upper face and viewer perception of uncanniness in virtual characters. I also take the opportunity to describe the origins and adaptive functions of facial expression in primates that help to set this experiment in context. NVC takes place while we speak, and the upper facial region plays an integral part in this process. An overview is provided of the purpose of NVC in humans and the associated roles of the eyebrows, eyelids and forehead and how they may work independently or alongside speech and other body and facial movements.

As a preeminent psychologist in the study of emotion types and their relation to human facial expression, in 2009 *Time* magazine listed Dr. Paul Ekman as one of the 100 most influential people in the world. As a result of nearly a lifetime's work, Ekman and his colleague Wallace Friesen devised a conceptual framework of the muscles involved in creating different expressions at differing intensity. This classification system of the facial muscles is referred to as the facial action coding system (FACS) and is commonly used as the blueprint of many modern facial modeling and animation software packages. I provide a retrospective of Ekman's work concerning the facial actions used in NVC and the implications of voluntary versus involuntary facial movements when detecting "false" expressions. Consideration is given to the consequences of a perceived lack of NVC in a human-like character's facial expression with and without speech. The main focus of this chapter and the arguments that I make about the possible cause of the uncanny in human-like virtual characters in this book are centered on a character's facial expression, especially the upper facial region. While this is an important factor in understanding the Uncanny Valley phenomenon, I acknowledge that facial expression should not be studied in isolation aside from other bodily movements and speech. The stimuli used in my experiments have featured vocalization narration, yet tilts of the head and gesture are integral parts of NVC in social communication. I am hopeful that the discussion of issues regarding facial expression and differing emotion types will work toward establishing a multidimensional model of the uncanny and help overcome uncanny facial expression of emotion.

4.1 UNIVERSAL EMOTIONS

Before Paul Ekman and Wallace Friesen (1978) could start to plan their classification system for facial expression, it was necessary to determine which facial expressions represent emotions that are common to us all. Since the early seventies, Ekman had been conducting studies to distinguish which facial expressions could be recognized at a universal level, irrespective of factors such as language, ethnicity, education, culture and age. Following several experiments with multiple participants from around the globe, six universally recognized basic emotions were established: anger, disgust, fear, happiness, sadness and surprise. Ekman (1992) stated that he had found strong, broad evidence of unique facial expressions to communicate these emotion types, among different cultures and levels of literacy. Furthermore, the recognition of these distinct facial expressions could be made whether an expression was made intentionally or if it was an unprompted, spontaneous reaction:

> This evidence is based not just on high agreement across literate and preliterate cultures in the labelling of what these expressions signal, but also from studies of the actual expression of emotions, both deliberate and spontaneous, and the association of expressions with social interactive contexts. (pp. 175–176)

Facial expression was key to establishing the basic emotions that were recognized by people who spoke different languages, were from varying levels of education and had different socioeconomic backgrounds. Researchers across various discourses including ethology, psychology and philosophy have long since pondered the significance of these different emotion types in humans and primates and why we are hard-wired to recognize facial expression regardless of language barriers. This multidisciplinary inquiry has resulted in a well-established theory: our innate ability to recognize and respond to another's facial expression serves as a primordial survival technique (Andrew, 1963; Blair, 2003; Darwin, 1872; Eibl-Eibesfeldt, 1970; Ekman 1972, 1992).

Ekman (1992) acknowledged that this antecedent skill helps us cope with fundamental life tasks both on a social and survival level. "Emotions are viewed as having evolved through their adaptive value in dealing with fundamental life-tasks. Each emotion has unique features: signal, physiology, and antecedent events" (p. 169). Whether we are at a group social gathering or in a one-on-one conversation, we are subconsciously tracking and reading other people's facial expression to understand how they are feeling and what they are likely to do or say next. This function allows us to respond appropriately to a person's emotional state and may be intensified if we interpret a more negative or hostile response from them. If someone is smiling and we perceive that they are happy, then we may feel at ease and wish to engage with that person to share their happiness. However, a flight-or-fight response may be evoked if we detect that a person is angry and about to start a fight. If we recognize that a person's facial expression shows disgust, then we may wonder what happened to evoke that emotion and, importantly, may look to avoid that potentially unpleasant or threatening situation ourselves (see, e.g., Blair 2003). Even if we perceive that we are on positive terms with someone that we have met for the first time, we may still be wary of any possible triggers of negative or hostile behavior from this unknown stranger. In this way, we benefit from a swift and accurate detection of how another is feeling, and we fundamentally rely on this adaptive skill to protect our own well-being.

Empirical evidence has revealed that we not only cognitively interpret and evaluate positive and negative emotions differently but also have a unique physical response to positive versus negative emotions (Darwin, 1872; Ekman, 1972, 1992; Johnson, Ekman and Friesen, 1990). A distinct physiology exists that involves emotion-specific sympathetic autonomic nervous system (ANS) patterns, dependent on the response required (Ekman, 1992; Johnson et al., 1990). Anxiety levels may be raised if we perceive more negative emotions such as anger, disgust, fear and sadness as we prepare to take action against potential harm. Hence, heightened ANS activity may occur for these negative emotions, though less so for more positive emotions such as happiness and surprise. If we are scared or detect fear in another, then an instinctive reaction is to get away from potential danger and blood rushes to the large skeletal muscles to support this survival response. Similarly for anger, blood rushes to the hands to prepare for a potential fight response (Ekman, 1992).

4.2 FACIAL ACTION CODING SYSTEM

We can instinctively recognize and interpret facial expression in humans and across different cultures, yet we still have limited understanding of the cognitive and physical processes that underlie our facial expressions and how these relate to experience of different emotions. Psychologists Ekman and Friesen wished to devise a method to help gain a better understanding of the physiognomic markers that represent emotion types. To do this, they devised the FACS (Ekman and Friesen, 1978) to taxonomize generic facial muscles involved in creating the six basic emotion types and at differencing levels of intensity. Facial muscles may act independently, though they typically act in unison to create different expressions. So that these facial actions could be measured and recorded systematically, Ekman and Friesen referred to these individual or combined movements as action units (AU). In the upper face, the corrugator supercilii is a small narrow muscle located above the eye that assists in moving the eyebrow downward. It was recorded that this muscle works in unison with an eye muscle called the depressor supercilii to depress the brows. This action pulls the eyebrows closer together and narrows the eye aperture. Deep wrinkles are formed between the brows, some vertical and others that may appear at a 45-degree angle. This corrugated furrow between the brows is often displayed when someone portrays a stern demeanor. Called the AU4, the brow lowerer, this facial movement is used to display more negative emotions such as fear, anger or sadness and may convey that a person is in deep thought, worried or concerned (Ekman and Friesen, 1978).

The frontalis muscle is situated as a layer across the front scalp that extends from the hairline to the brows and consists of the inner and outer frontalis muscles. Action Unit 1 (AU1), referred to as the inner brow raiser, describes the changes in appearance that the inner frontalis muscle creates in the upper face. When contracted, this muscle pulls the scalp back and lifts the inner corner of the eyebrows. As a visual result of this action, wrinkles can appear (and existing wrinkles can deepen) in the center of the forehead. Action unit two (AU2), named the outer brow raiser, describes the contraction of the outer frontalis muscle that lifts the outer part of the eyebrow. This creates lateral wrinkles in the outer sides of the forehead and the eyebrows are raised forming an arch shape. AUs 1 and 2 often work together to open the eye area and increase the visual field (Ekman, 1979; Ekman and Friesen, 1978). As well as a visual task to help focus on distant objects, this combined muscle movement can be perceived as a signal of surprise and interest. Ekman and Friesen (1978) systematically determined that AU1, AU2 and AU4 are involved either individually or in combined actions to create one of the several defined brow movements evident in the six basic emotions (see also Ekman, 1979).

In addition to being used extensively across the fields of psychology and medicine, games and animation companies have utilized this database of action units as a reference tool for creating facial expression of emotion in virtual characters. The Source Software Development Kit (SDK) was developed by Valve in 2008 and allows designers to create games for the Source video game engine. *FacePoser* (Valve, 2008) is a facial expression tool within Source SDK that allows a designer to create static and animated facial expressions for game characters. Each action unit is assigned to an interactive flex slider bar so that the character's facial muscles can be controlled as individual parameters or to work in unison to create the desired expression. Expressions can be created at differing levels of intensity by moving the slider bar, shifting to the right to increase intensity and then backward to the left to decrease intensity. To create a frown expression, the flex slider bar, known as the lowerer, can be manipulated to simulate AU4 in the upper face. It could be selected along with other AUs in the lower face required to represent a facial expression of anger in a character. This technique can be used to generate characters' facial expressions both in prerecorded and real-time game footage. In addition to characters with a highly human-like appearance, FACS is also utilized by animators for generating facial expression for anthropomorphic characters. In her article "Written All over Your Face," journalist Debbie Hampton (2011) recognized the powerful synergy between FACS and animation and how it has allowed those creating characters the ability to better read and understand expression. "Computer animators even employed it at Pixar when making *Toy Story* and at Dream Works for making *Shrek*. Those who have mastered the system and information yielded gain extraordinary insight into reading faces" (p. 1).

4.3 NONVERBAL COMMUNICATION

Many place the importance of facial expression as a commutation tool above that of speech (and other vocalization narration) and body movements (Darwin, 1872; Ekman, 2004; Izard, 1971). Support for this proposal can be found in an experiment that psychologist Carroll E. Izard conducted in 1971 using rhesus monkeys. These particular primates were used because their communication methods may be compared with the evolutionary development of human communication based on the ancestral phylogenetic relationship between rhesus monkeys and humans (Darwin, 1872; Izard, 1971). The facial nerves of a group of monkeys were cut so that they were unable to make facial movements (Izard, 1971). This group was then placed with a group of monkeys with normal facial movement. When the two groups intermingled, the monkeys with disabled facial movement used full-body gestures as an alternative means of communication with the other monkeys. These movements escalated in aggression to the point that some monkeys with normal facial expression were attacked by the monkeys with no facial expression. The monkeys in the experimental group may have been aware of their lack of ability to use facial expression and grew frustrated at not being able to communicate this way with the other monkeys. Without the use of facial expression, they risked not being fully understood by the other monkeys. In this way they may have resorted to aggression at the sheer frustration of not being understood. Furthermore, the monkeys with disabled facial movement may have felt threatened by the other monkeys with full facial expressions. Without the ability to communicate facially, they may have sensed that they were at a distinct disadvantage to this other group and physically attacked them as a defense mechanism (Izard, 1971).

In humans, our behavior is instilled with meaning, and an observer and/or interactant can gain crucial information about a person's emotional state based on their face and body movements (Ekman, 1965, 1979, 2004; Ekman and Friesen, 1969, 1978). In addition to facial expression, NVC can be conveyed in various ways, including the posture of a person's body, their hand gestures, and the volume, delivery style and tone of a person's speech. NVC can negate the need for speech in that an observer can understand that a person is angry by a narrowing of their eyes and a shaking of their fist (Ekman, 1965, 2004). Furthermore, NVC made voluntarily (so that the person is consciously aware of it) or involuntarily (as a subconscious process) can reveal much more about that person than just their emotive state. In an emotion recognition study (predating the publication of FACS in 1978), Ekman and Friesen (1969) discovered that participants not only were able to identify an individual's emotive state by observing their face and body movements but could also make judgments about the individual's personality, social skills and attitude via NVC. Even those participants with no specialized training in emotion recognition were able to make accurate decisions that matched people's independent assessments of their own personality traits, social skills and attitudes.

In addition to the pursuit of identifying and classifying basic emotion types, Ekman also wished to pursue the complex interrelationships between voluntary and involuntary facial movements (see Ekman, 2003; Ekman and Friesen, 1969, 1982)—in other words, how we differentiate between those facial expressions that we contrive and are aware that we are making and those that occur spontaneously as an automatic response to a situation. As a way toward classifying these voluntary versus involuntary facial movements, by 1979 Ekman had established two different types of facial NVC: conversational actions and emotional expressions. He noted that both types of these facial social signals can occur during speech, but at different points in a conversation. Conversational actions typically occur as a voluntary action and are used either before or at the same time a word is spoken. Involuntary actions also occur during speech as emotional expressions to semantically accentuate the meaning of a word. During conversation, the listener may also use emotional expressions as a way to express their opinion as another speaks, for example, to show that they either agree or disagree with the speaker (Ekman, 1979).

Of particular significance to the theories that I present in this book on the cause of the uncanny in virtual characters, Ekman (1979) suggested that the brows are used more frequently as voluntary conversational actions than other parts of the face. The brows are used in this way because they are contrastive and may present either positive or negative emotions. Brow movements are also among the easiest to perform compared with other facial (or body) movements because the lips and lower facial regions may be involved in speech. Throughout his studies, Ekman (2004; see also Ekman and Friesen 1969) recognized the importance and capacity of the upper facial region to transmit messages that may be interpreted globally and intuitively. As such, two further categories of nonverbal signals presented during speech were defined as emblems and illustrators, both of which incorporate movement in the upper facial region (Ekman, 2004; Ekman and Friesen, 1969). As the name suggests, illustrators describe movements that help to illustrate speech in real time. Tied to particular words as they are spoken, brow lowering and raising serve to augment what is being said. Additional emphasis may be added to a more negative word if it is spoken with a lowered brow (AU4) because this facial movement is associated with more negative emotions such as anger, disgust, fear and sadness. Ekman coined such brow movement as baton accents (Ekman, 2004, p. 41), which are distinctly related to the content of the word being said. For example, a frown expression may be concomitant with the word "no" to accentuate the anger and defiance underlying the spoken refusal. More positive words such as "easy, light, good, etc." (p. 42) may be baton accented by AUs 1 and 2, which raise the brows. This action is associated with more positive emotions such as happiness and surprise and, used in this way, may help to indicate that one is interested in and enthusiastic about what they are saying (or listening to).

The importance of context and accurate facial expression in the upper face region is prevalent here for human-like virtual characters during speech. Raised brows may be indicative of the word *excellent* said to express delight at a given outcome. However, if this word is spoken with still brows, or even lowered brows, in a human-like virtual character due to a lack of detail or accuracy in animating the upper facial region, then the word may be perceived as an expression of sarcasm and that the character is more dissatisfied than pleased with a particular outcome. Hence, if the designer has created this effect unwittingly for a character due to a lack of accuracy in the character's upper face (i.e., they intended that the character should be pleased, but the character's upper facial movements depict a lack of enthusiasm), then a viewer may perceive that the character is not being completely honest and does not really mean what they are saying. The implications of a suspicion of a false emotion being presented in a virtual character due to inadequate NVC and the significance of emblems as a possible attempt to deceive are discussed in the next section.

4.4 FALSE OR FABRICATED EMOTION

Charles Darwin (1872) suggested that despite people's best efforts to try to conceal or hide an emotion, their facial expression often revealed how they were really feeling. Interestingly, Darwin also proposed that people could detect when they were being presented with a false expression in an attempt to appease another, despite that emotion not actually being felt. Building on Darwin's work and the findings from emotion recognition studies prior to the publication of FACS Ekman and Friesen (1969) grew increasingly aware and curious of perception of spontaneous versus deliberate facial expression and the detection of involuntary, transient NVC to suggest possible deceit. Importantly, they realized that NVC could reveal much more about a person's true emotional state than their spoken words and other verbal cues.

> Nonverbal behavior provides valuable information when we can't trust what we are told in words, either because the speaker is purposefully trying to deceive us, or because he has blocked or repressed information we want … (Ekman and Friesen, 1969, p. 52)

Despite a person's best efforts to deceive another by consciously controlling their facial expression and speech, some NVC is uncontrollable and may leak clues as to how that person is really feeling and that they may be attempting to deceive (Ekman and Friesen, 1969; Ekman, 2004). While people may be able to stop themselves from clenching their fists to hide that they are angry, they may not be able to prevent the frown expression that appeared on their face in response to their anger. People who have achieved second place in a competition may smile as they congratulate the winner but may be unable to prevent their lowered brow, which signals their disappointment. Even though they may think that they are presenting a happy face, this NVC in their upper face reveals how they are really feeling not to have won. Emblems help reveal further how people may be unable to control facial movements that unmask their true thoughts and feelings (Ekman and Friesen, 1969; Ekman, 2004).

The majority of emblems are made intentionally, and a person is often aware that they are using their face or body to send a given signal to the listener (Ekman and Friesen, 1969; Ekman, 2004). As such, emblems are typically performed in the "presentation position" (Ekman, 2004, p. 40) when the speaker is facing others. A person may intentionally raise their brows to signal that they are surprised and tilt their head backward to accentuate their surprised response. However, it was revealed that some emblems can occur unintentionally (Ekman and Friesen, 1969; Ekman, 2004). A person may be unaware that they have made a particular gesture as they may not have wished to share this information with others. These inadvertent movements that may give away how a person truly feels were referred to as emblematic fragments and were likened to a verbal slip of the tongue. "Like verbal slips, emblematic fragments may reveal repressed information, or deliberately suppressed information" (Ekman, 2004, p. 40). These momentary, uncontrolled facial expressions may be as obvious to the viewer as when a person accidentally speaks their mind and reveals information they would rather not have disclosed.

Ekman (2001, 2003) conducted further studies to investigate how involuntary facial expressions may be used to detect deception—in other words, how a face could be decoded to tell when a person was telling lies. This liar detection system was based upon recognizing involuntary movements that could not be made without spontaneous emotional response. Hence, these specific movements for a particular emotion were identified as reliable signals of a person's true emotional state (ibid.). As well as a lowered brow (activated by AU4, brow lowerer), anger requires that AU24, called the lip presser, be active in the lower facial region for people to perceive that anger is actually felt. AU24 is seen when the orbicularis oris muscle that surrounds the mouth contracts, pressing the lips tightly together. The unique combination of AUs 1+2+4 is an involuntary, uncontrolled action that signals genuine fear is felt (Ekman, 2001, 2003). Given that the adaptive function of fear is an anticipatory response of distress, Ekman (1979) provided an explanation for the origin of this reliable facial movement for fear. "… It might make sense to explain 1+2+4 as a movement relevant to attention and increased visual input (1+2), occasioned by a novel object, with 4 providing the clue as to what is anticipated, a distress experience" (p. 198). AUs 1 and 2 allow for increased visual field and attention, while AU4 signals anticipation of panic and distress. If a person tries to communicate a particular emotional state without evidence of these reliable expressions, then one may question the authenticity of the portrayed emotion with the possibility that the person is lying (Ekman, 2001, 2003). With regards to human-like virtual characters, if a character's facial expression is not modeled correctly and a character's brows remain or become lowered when they should not be, then a viewer may not be convinced that the character is as calm or content as their voice and situation would otherwise lend them to be. As I discuss later in this chapter, rather than too many inadvertent or accidental facial expressions occurring to raise suspicion of deceit, it may be that a distinct lack of these reliable, involuntary facial expressions raises suspicion of false emotion and thus provokes the uncanny.

4.5 THE EFFECT OF EMOTION TYPE ON UNCANNINESS

Ekman's work (and the work of others) stressed the importance of NVC in the face, especially in the upper face, as the primary source of detecting the affective state of an individual. Furthermore, the perceiver relied on this NVC so that they could respond promptly to the sender and avoid potential harm if that person was communicating more negative emotions. Based on this, in 2011 I designed an experiment (authored with Grimshaw, Williams and Abdel-Nabi) to assess the significance of different emotion types on perceived uncanniness in human-like virtual characters and the implications of a lack of upper facial movement when communicating emotion types. I predicted that survival-related emotions such as anger, fear, sadness and disgust (Ekman, 1979)—which could be considered as signals of a potential threat, harm or distress—would be rated as more uncanny when presented in near human-like, virtual characters than more positive emotions such as happiness and surprise (Tinwell et al., 2011). I also expected that uncanniness would be intensified when facial expression was limited in the upper facial region as the ability to recognize the character's emotional state may be impaired. The stimuli for this experiment consisted of headshot videos of three groups: (1) human, (2) full, and (3) lack.* The human group was an 18-year-old male actor who was instructed to present prosodically congruent utterances to match facial expressions of anger, disgust, fear, happiness, sadness and surprise. To create the full group, the facial expressions and speech from the human videos were replicated onto Barney, a male character with a human-like appearance from *Half-Life 2* (Valve, 2008). Then, for the lack group the videos from the full group were replicated, but this time movement was disabled in the upper facial region including the eyelids, eyebrows and forehead. For example, AU4 was disabled for anger and AUs 1, 2 and 4 for fear. To help visualize these different facial expressions of emotion, I have provided examples of a male and female human and male and female virtual characters in the full and lack states as shown in Figures 4.1 through 4.6. These were not the original stimuli used in this experiment, but these images help to demonstrate how human facial expression of emotion can be replicated in human-like virtual characters with full facial animation and how a lack of movement in the upper face can reduce the overall communication for each emotion. The figures show examples of human, full and lack characters expressing fear (Figure 4.1), surprise (Figure 4.2), sadness (Figure 4.3), disgust (Figure 4.4), anger (Figure 4.5) and happiness (Figure 4.6).

The 116 participants who took part in the study rated the fully animated virtual characters as less familiar and human-like (more uncanny) than the human videos. Overall, partially animated characters in the lack group received significantly higher uncanniness ratings than those fully animated virtual characters (Tinwell et al., 2011). As was predicted, characters were perceived as more uncanny when NVC was disabled in the upper face as this lack of emotional expressivity had obscured and reduced the salience of an emotion in a character. We rely on a swift and accurate detection of another's emotion so that we may predict their impending behavior. Hence, the viewer may have been confused by the lack characters' emotional state and the future actions of those partially animated characters (Tinwell et al., 2011). When verbal and nonverbal behavior are combined, a viewer expects

---

* For a full description of the methodology, design, procedure and results of this study please see Tinwell et al. (2011).

congruence between the two to confirm another's affective state (Tinwell et al., 2011). Without this congruence between verbal and nonverbal behavior, the viewer may be concerned as they do not know how to behave in response to that character and how that character will behave toward them.

> Any observed incongruence alerts people to oddness and the possibility of unpredictability of behavior which is alarming (even distressing and scary) as it may present a potential threat to personal safety. Hence, the sensation of uncanniness may serve to act as a sign of unpredictability and danger. (p. 746)

As such, it was proposed that these results may help us gain further insight as to the function of the uncanny. Specifically, experience of the uncanny is a warning to alert us to possible unpredictable, threatening behavior in a character (Tinwell et al., 2011). Interestingly, the magnitude of this increased uncanniness in the lack group varied depending on which emotion was being presented. In the lack condition, perception of uncanniness was strongest for the emotions fear, sadness, disgust and surprise, though less so for anger and happiness. The possible reasons as to why emotions with distinctly different adaptive functions (social versus survival) are more prone to the uncanny than others are discussed next.

4.5.1 Uncanny Emotion: Fear and Surprise

When compared with the other emotions, fear (see Figure 4.1) was rated as most uncanny in characters with a lack of upper facial movement (Tinwell et al., 2011). Given that fear requires only small facial movements compared with the other emotions, it was suggested that a character without movement in its upper face may have resembled a corpse when communicating this emotion. In this way, the character may have represented a reminder of death's omen and be placed alongside the corpses that Mori (1970/2012) initially placed in the Uncanny Valley (Tinwell et al., 2011). This possible reminder of death not only may disturb the viewer but also may compromise our innate ability to recognize and respond to adverse emotions in others (Tinwell et al., 2011).

The psychologist Robert James R. Blair stated in his research on aversive emotion types that we have an instinctive reaction to avoid fear. "Fearful faces have been seen as aversive unconditioned stimuli that rapidly convey information to others that a novel stimulus is aversive and should be avoided" (Blair, 2005, p. 702). Therefore, aberrant facial expression in

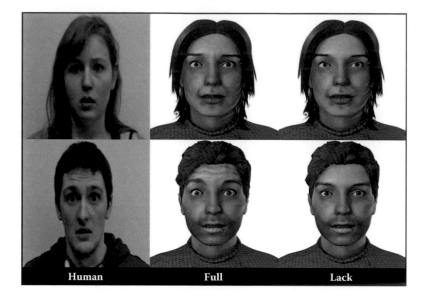

FIGURE 4.1 An example of male (bottom) and female (top) humans (Human) and human-like virtual characters with full facial animation (Full) and without movement in the upper face (Lack) presenting fear, created by animator Dr. Robin J. S. Sloan.

the lack character for the "aversion reinforcement" (Tinwell et al., 2011, p. 747) of the emotion fear may have increased uncanniness as people are gravely aware of the possible harmful ramifications of misunderstanding this negative emotion. Thus, the character's ambiguous facial expression may have triggered a fight-or-flight response (Tinwell et al., 2011, p. 747) in the viewer to avoid the potential repercussions of misinterpreting fear. The uncanny was strongest for fear as the viewer was faced with the dilemma of not being able to understand and respond to a possible negative, fearful emotional state in another. Without evidence of the reliable AUs 1, 2 and 4 for fear (Ekman, 2001, 2003), the participants may have been doubtful whether the character was attempting to communicate fear or not. As such, in addition to the character being perceived as strange due to their aberrant facial expression, anxiety may have been raised in the viewer as they simply did not know how to react to that character and what the potential negative consequences of misreading that character's emotional state may be: Should they prepare for a fight-or-flight response or not?

Perception of the uncanny in the human-like virtual character increased significantly between the full and lack states when surprise was presented (see Figure 4.2). This finding was a surprise in itself given that accurate

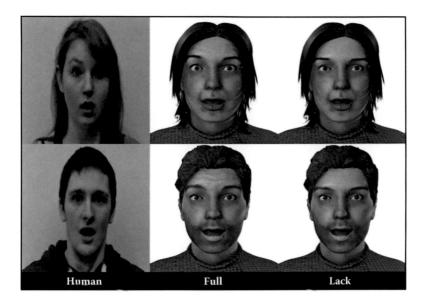

FIGURE 4.2   An example of male (bottom) and female (top) humans (Human) and human-like virtual characters with full facial animation (Full) and without movement in the upper face (Lack) presenting surprise, created by animator Dr. Robin J. S. Sloan.

detection of surprise in others may not be as crucial to one's survival as the ability to accurately detect other more negative emotions, such as fear or anger. Ekman's (2003) work has repeatedly shown that in humans surprise is commonly mistaken with fear due to their similar facial actions. Therefore, it was suggested that as surprise (see Figure 4.2) is commonly mistaken with fear in humans, the same characteristics attributed to fear in the virtual character lack condition may be applicable to the emotion surprise (Tinwell et al., 2011).

In real life, surprise may be more distinguishable from fear due to the events that preceded a surprise reaction. For example, surprise is likely to occur walking into a room full of people cheering to celebrate your birthday when you thought that they had forgotten or on discovering that you have passed an exam that you had otherwise expected to fail. These events may lead to less ambiguity as to whether someone is presenting a surprised or fearful expression as their expression is contextualised by the situation. The videos in this experiment comprised just a headshot against a monotone background, with no other stimuli or objects to set a scene or character's behavior in context. Therefore, it was presumed that some of the characteristics that enhance the uncanny for fear may be evident for the emotion surprise (Tinwell et al., 2011). This may help explain why uncanniness ratings for surprise were more typical of negatively valenced emotions such as fear, sadness and disgust and less likely to resemble a more positive valenced emotion.

4.5.2 Sadness and Anthropomorphism

Our ability to anthropomorphize characters may have been evident when it came to rating fully and partially animated characters for the emotion sadness. Even though removing facial expressivity in the upper facial region significantly increased the uncanny for this emotion, sadness (see Figure 4.3) was still rated as the most familiar (least strange) and human-like emotion when presented in the virtual character in both the full and lack states (Tinwell et al., 2011). When we encounter a human-like character expressing sadness, we may wonder what caused them to be sad and express sympathy toward them (Tinwell et al., 2011). This sympathy may extend to a level of empathy for that character in that we wish to help them and prevent them from being sad.

In humans, a distinctive characteristic of sadness is a droopy mouth shape. This movement is created by a muscle that is attached to the lower

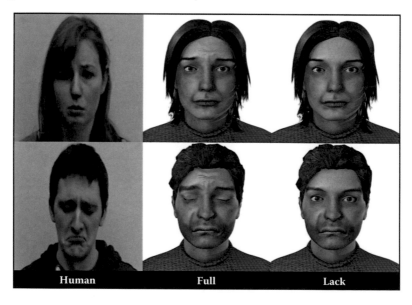

FIGURE 4.3 An example of male (bottom) and female (top) humans (Human) and human-like virtual characters with full facial animation (Full) and without movement in the upper face (Lack) presenting sadness, created by the animator Dr. Robin J. S. Sloan.

jaw. Known as the depressor anguli oris muscle, it pulls the corners of the mouth downward. This action is described by AU 15, lip corner depressor, and is simulated in various facial animation software, thus allowing animators to create a sad expression in the lower face for virtual characters. AUs 1 and 4 are also involved in sadness in the upper face to form slanted eyebrows, a frown and the associated wrinkles that these movements create. Even when a character may lack sufficient detail in the upper face, the depressed, downward corners of the mouth may be enough to compensate. The viewer is still able to understand this sullen expression and that the character may be feeling helpless or defeated in some way. A sense of sympathy may override a sense of strangeness for the character as the viewer may be less discerning (and more tolerant of) abnormal facial expression due to the character's distressed state (Tinwell et al., 2011).

4.5.3 Disgust, Revulsion and the Nose Wrinkler Action

It is widely accepted that the facial expression of disgust (see Figure 4.4) serves the adaptive function of a warning signal to others to be alerted to the possibility of potential infection or encountering a distasteful object (Blair, 2003; Darwin, 1872; Rozin, Haidt and McCauley, 1993). One may

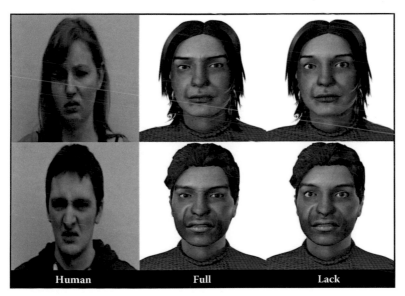

FIGURE 4.4   An example of male (bottom) and female (top) humans (Human) and human-like virtual characters with full facial animation (Full) and without movement in the upper face (Lack) presenting disgust, created by animator Dr. Robin J. S. Sloan.

withdraw from a person or object that they regard as revolting or unclean to protect one's own health and well-being.

Perception of the uncanny increased significantly going from human to full to the lack state for disgust, providing support for my initial theory that people would be much more uncomfortable when this emotion was presented in virtual characters with aberrant facial expression (see Tinwell et al., 2011). A viewer may have experienced confusion when presented with the lack disgust state with questions arising such as: What is this character attempting to communicate? Have they encountered something that has repulsed them or not? If so, how should I react, and am I at potential risk of infection or disease? In this way, a perceived delay in avoidance strategies caused by aberrant facial expression in the character may make the viewer feel distinctly uncomfortable and elicit the uncanny.

While it was considered that the unsettling reaction that disgust can elicit—as a warning to avoid a repugnant object or unpleasant (even harmful) situation—caused this uncanny response, current limitations in facial animation software may also have contributed to uncanniness (Tinwell et al., 2011). In humans, a curling of the upper lip and acute wrinkling of the upper nose are two distinctive facial features of disgust (Ekman and Friesen, 1978; Ekman, 1992). This expression is created when the levator labii superioris alaeque nasi muscle, which runs symmetrically down either side of the top half of the nose to the upper lip, contracts. This action lifts the upper lip and dilates the nostrils, thus compressing the nose and causing wrinkles to form as if the person was snarling. Ekman and Friesen (1978) defined this action as AU9, nose wrinkler, though the skin resolution for a virtual character may not be adequate to effectively simulate this action when used in facial animation software. The lower pixel count available, especially in video game engines, to achieve detailed and complex folds in the skin may have been limited in the virtual character used in this experiment. This reduced graphical fidelity in the skin texture to create the nose wrinkler action may have compromised the effective communication of disgust, especially when movement was restricted in the upper face. A perceived mismatch may have occurred between the curled upper lip shown in the lower face and a lack of movement in the upper face. Without a clear representation of this most distinctive nose wrinkler movement involved in disgust, the character's expression appeared abnormal and bizarre, thus exaggerating the uncanny.

It was put forward that disgust may remain as an uncanny emotion until facial animation software can effectively convey this particular characteristic. "Until modifications are made to improve the graphical realism for this facial movement, disgust may continue to be perceived as uncanny regardless of modifications made to other parts of the face" (Tinwell et al., 2011, p. 747). As discussed in later chapters of this book, performance capture techniques such as motion scanning may allow for a more accurate depiction of folds and wrinkles in the skin to aid the effective communication of disgust (and other emotions) in prerecorded animation. However, for facial expression created in real-time in a gaming environment, a comparatively lower pixel (and polygon) count achievable for skin textures may serve as an ongoing limitation for uncanny disgust.

4.5.4 Our Perceptual Advantage to Anger

As one of the most important survival-related emotions, facial expressions of anger (see Figure 4.5) signal a warning of hostility and possible violence in another. Similar to fear, the detection of anger is crucial in one's survival instincts, in that we have primal impulses to fight or escape when in the presence of anger. The results from my experiment showed that participants rated videos of the human expressing anger highest for perceived familiarly and human-likeness (Tinwell et al., 2011). This highlights our

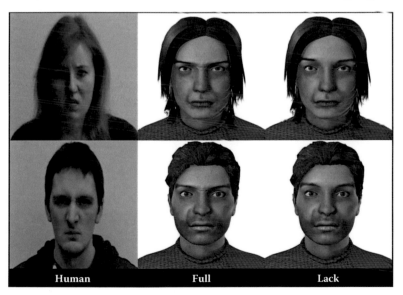

FIGURE 4.5   An example of male (bottom) and female (top) humans (Human) and human-like virtual characters with full facial animation (Full) and without movement in the upper face (Lack) presenting anger, created by animator Dr. Robin J. S. Sloan.

intrinsic ability to recognize a clear communication of anger so that we can respond accordingly to prevent potential harm or threat.

Even when anger was displayed in the lack character without upper facial movement in a more primitive, less human-like state, anger was rated as no more uncanny than when presented in the fully animated human-like character. As a more negatively valenced emotion, it was expected that characters displaying this emotion without upper facial movement would be regarded as significantly more uncanny than those fully animated characters due to a possible delay in detecting such an important warning of a potential threat (Tinwell et al., 2011). Yet no significant difference was found in ratings for uncanniness between the full and lack states presenting anger (Tinwell et al., 2011). In other words, the uncanny was not increased any further for human-like characters when they presented anger without upper facial movement than when this emotion was communicated using both the upper and lower face. A lowered brow and intense frown expression is a principal feature of anger (Darwin, 1872; Ekman, 2004; Ekman and Friesen, 1978) so it was expected that without this vital sign uncanniness would be intensified as the viewer may panic as to how they should react to the character. However, the results that we collected challenged this assumption. I therefore attributed this finding to our instinctive ability to recognize anger in another as a primitive survival mechanism, despite aberrant facial expression (Tinwell et al., 2011).

It has been found that in addition to upper facial movement, people also rely on a tightening of the lips to signal that genuine anger is felt (Ekman, 2001, 2003). This involuntary action (AU24) created by the muscle that surrounds the mouth provides a reliable visual sign for anger. This may help explain why the uncanny was less noticeable in the human-like character when he presented anger without upper facial movement. His pressed lips both during and between speech clarified the authenticity of this emotion (Tinwell et al., 2011). The exaggerated tightening of the lips still sent a clear warning to participants that anger was felt. In this way, the viewer may still be able to detect anger in human-like virtual characters even with no upper facial expression by relying on lower facial expression in addition to articulation and prosody.

4.5.5 Happiness and Uncanny False Smiles

As I had predicted, participants were less sensitive to the uncanny when the human-like character presented the emotion happiness (see Figure 4.6) with a lack of upper facial movement when compared with the

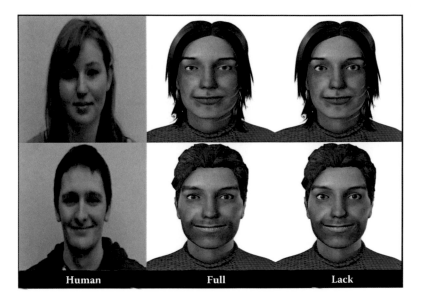

FIGURE 4.6   An example of male (bottom) and female (top) humans (Human) and human-like virtual characters with full facial animation (Full) and without movement in the upper face (Lack) presenting happiness, created by animator Dr. Robin J. S. Sloan.

other six basic emotions (Tinwell et al. 2011). Surprisingly, the results also revealed that happiness was rated as the least familiar and human-like (most uncanny) emotion when the human-like character presented this emotion with full facial expression. This finding was unexpected due to happiness being a more positive emotion. The human-like character was also perceived as less uncanny when presenting happiness in the lack state when compared with the full state for happiness. Building on previous work by physiologists and psychologists to determine genuine versus false emotions, I suggested that participants may have been suspicious of being presented with a false smile (Ekman and Friesen, 1982).

The phenomenon of false smiles, and the reason that humans are naturally so good at detecting them, can be traced back to the work of French physiologist Guillaume-Benjamin-Amand Duchenne (1862). The zygomaticus major muscles extend from each corner of the mouth to the cheekbones on either side of the face. When each muscle is contracted, they pull the corners of the mouth upward to create a smile shape, and dimples may occur in the cheeks. In 1978, Ekman and Friesen classified this action as AU12, lip corner puller. Importantly, Duchenne (1862) identified that movement of the zygomaticus major muscle can be controlled

FIGURE 4.7 Crow's feet wrinkles and bulges that appear around the eye area when spontaneous and genuine happiness is felt.

voluntarily to create an upward curl of the lips to simulate a smile. In this way, a person may create a smile expression even though they may not be feeling happy. However, movement of the orbicularis oculi muscle that raises the cheek and narrows the eye aperture, gathering skin inward around the eye socket during a smile, occurs involuntarily in response to experiencing spontaneous, felt happiness (Duchenne, 1862; Ekman, 2001, 2003; Ekman and Friesen, 1982). This movement, defined as AU6, cheek raiser (Ekman and Friesen, 1978), creates two distinctive physiognomic markers around the eye area. Bulges appear below the eye socket as pockets of fat are compressed, and crow's feet wrinkles (Ekman and Friesen, 1982, p. 243) appear beside the eyes as the skin gathers inward toward the eye socket (see Figure 4.7). Both these bulges and crow's feet wrinkles signal to the viewer that genuine happiness is felt.

Therefore, AU6 may be used as a measurable difference between felt and false smiles in that this action is evident in felt smiles and absent in false smiles. The low-polygon counts and number of pixels achievable in the skin texture around the eye area for the human-like character may have served as a significant obstacle in simulating a realistic smile when he was presented in the fully animated state. As such, this low resolution in the eye area did not adequately convey the bags and creases below the eye or the crow's feet around the eyes that would otherwise be visible in humans. Furthermore, any movement that occurred in the brow and eye area as this human-like character smiled, but without the appearance of crow's feet wrinkles around the eyes, may have suggested incongruence in his behavior with a distinct mismatch between his upper and lower facial expressions.

If a person voluntarily creates a smile shape in the lower face in an attempt to communicate happiness when they are actually experiencing more negative emotions, AU12 may be used in conjunction with muscles in the upper face that resemble characteristics of anger, fear, sadness or disgust (see, e.g., Darwin, 1872; Duchenne, 1862; Ekman, 2003; Ekman and Friesen, 1982; Frank, Ekman and Friesen, 1993). Hence, the perceiver may detect a false smile if the upper face presents facial expressions associated with more negative emotions, contradictory to those shown in the lower face. Therefore, if a human-like, virtual character presents a smile, then this may not be sufficient to convince the viewer that a positive emotion is felt if that character's upper face is not modeled correctly. The viewer may be able to detect a reliable expression of fear or trepidation in the character's upper face that reveals a leakage of hidden, more negative emotion. This may explain why the uncanny was actually reduced for the human-like character when he expressed happiness with no upper facial movement when compared with the fully animated state (Tinwell et al., 2011). Without the clear and appropriate appearance of crow's feet wrinkles and bulges around the human-like character's eyes in the full state, participants may have been confused as the character's upper face appeared to show more negative emotions in contrast to the more positive emotion portrayed by a smile shape.

A recent example of this uncanny effect due to a lack of detail and expression around the eye area is evident in the computer-generated (CG) simulation of actress Audrey Hepburn* in the *Galaxy Chauffeur* (Kleinman, 2013) commercial. Animators at Framestore worked with director Daniel Kleinman to create an advertisement in which a CG Audrey Hepburn was brought back to life by capturing the performance of an Audrey Hepburn lookalike. Set in Italy's Amalfi coast, Audrey Hepburn was invited to escape a crowded, broken-down bus and take a ride with a luxury chauffeur while enjoying Galaxy chocolate. Specifically, Audrey Hepburn's face shows a distinct lack of emotional expressivity, and there are points when her smile may be perceived as false to suggest that she may be hiding more negative emotions. There is a mismatch between a lack of NVC in her upper face and her wide smile in the lower face. A lack of the crow's feet wrinkles and bulges that appear around the eyes to signal genuine happiness risk her smile being perceived as false. Hence, the viewer may be confused if this character is actually happy and genuinely enjoying the moment, or if she is putting on a brave face and putting up with her current situation, that works against the purpose of this commercial. Even though this character's mouth is smiling, her eyes are not engaged in the smile. A viewer

---

would expect that she should be happy and excited, being swept away in the moment and enjoying the luxury travel and chocolate. However, her smile appears rather stoic as she nods to the handsome chauffeur, and as she eats the chocolate there is a lack of the involuntary visual cues that one would detect to signal true pleasure and happiness.

As William Bartlett, Framestore's VFX supervisor, explained, the project required a full CG face replacement of the actress filmed for the animation "to bring the benchmark of beauty, Audrey Hepburn, back to the screen" (Bartlett, 2013, p. 1). The Framestore team used innovative facial-scanning techniques and also FACS to record and create the human-like character's facial muscle movements. No less than 70 facial actions were prerecorded from the actress to help build the CG Audrey Hepburn (Bartlett, 2013). In addition to this, a powerful Arnold (Solid Angle, 1999) rendering system was employed to accurately trace light and enhance the translucency and quality of the skin texture. Yet, despite this, the CG Audrey Hepburn still failed to portray authentically believable emotion that matched what was happening to her in the context of the advertisement. Accordingly, and in support of the observations that I have made for this character, Bartlett remarked that "the biggest challenges in recreating an authentic and unmistakable Audrey Hepburn proved to be the eyes and smile" (p. 1). I question why this should occur given the advanced technology, high skill set and expertise applied to this character's facial expression? On contemplation, various factors could have contributed to this. It may have been a lack of genuine happiness captured in the actresses' performance itself; given that it was a staged and acted performance in the recording. This may have resulted in a lack of spontaneous happiness and a more fabricated, contrived emotion from the real-life actress. The level of detail achievable in the skin texture may still have been inadequate to fully depict the fine details around the eye area for this CG character. Another contributing factor may have been the emphasis placed on retaining Audrey Hepburn's iconic beauty.

If designers are modeling a female protagonist character who is supposed to represent a beautiful heroine in a game or animation, then facial cues such as bulges and crow's feet wrinkles may be regarded as unattractive, unnecessary flaws rather than important and crucial signals. The designer may be reluctant to include such details in case they represent a more ugly appearance and imperfections in her flawless skin. A designer may even intentionally erase these less attractive wrinkles and folds in the skin to retain that character's perfect appearance. As well as detecting this uncanny false happiness in the male character, Barney (that was included in my experiment), this problem may be worsened in female characters, especially in the video games industry. Traditionally in the games industry (and popular culture), females are designed with a highly attractive appearance and few physical imperfections. Despite their perceived beauty, such characters may remain uncanny when expressing the emotion happiness due to a lack of detail around the eye area. Their behavior may be regarded as unnatural and nonhuman-like, with implications of the Botox effect that Clive Thompson (2004) observed in female video game characters. If these crucial details around the eye area are missing in a female protagonist and the viewer is suspicious that they are being presented with a false smile, then this may make the character appear less sincere and trustworthy. Ironically, attempts to improve the attractiveness of empathetic, female human-like characters by avoiding imperfections such as wrinkles may result in a character being perceived as less attractive. Instead, more negative personality traits may be associated with that character, and the viewer may begin to dislike them. If continually presented with potential deception clues, the viewer may be less trusting of that character. Judged as capable of lies of commission or omission, the viewer is less able to establish a sense of rapport or familiarity with them. The uncanny effect is instantiated as the viewer is wary of deceit or a false impression. Such tactics may work to the advantage of antipathetic characters such as criminals or monsters intended to be regarded as insincere and untrustworthy. However, it could work against protagonist characters intended to be perceived as trustworthy and of a happy, positive demeanor. We may doubt the character's intentions and ultimately fail to connect with them. The consequences of more negative, antisocial personality traits being associated with uncanny human-like virtual characters due to their aberrant facial expression is discussed further in the next chapter.

REFERENCES

Alone in the Dark (2009) [Computer game], Atari Interactive Inc. (Developer), Sunnyvale, CA: Atari Group.

Andrew, R. J. (1963) "Evolution of facial expression," *Science*, vol. 142, pp. 1034–1041.

Bartlett, W. (2013) "Galaxy chauffeur," *Framestore*. Retrieved January 30, 2014, from http://www.framestore.com/work/galaxy-choose-silk-chauffeur.

Blair, R. J. R. (2003) "Facial expressions, their communicatory functions and neurocognitive substrates," *Philosophical Transactions of the Royal Society*, vol. 358, pp. 561–572.

Blair, R. J. R. (2005) "Responding to the emotions of others: Dissociating forms of empathy through the study of typical and psychiatric populations," *Consciousness and Cognition*, vol. 14, pp. 698–718.

Darwin, C. (1965) *The Expression of the Emotions in Man and Animals*. Chicago: University of Chicago Press. (Original work published in 1872.)

Duchenne, B. (1862) *Mecanisme de la Physionomie Humaine ou Analyse Electrophysiologique de l'Expression des Passions*. Paris: Bailliere.

Eibl-Eibesfeldt, I. (1970) *Ethology, the Biology of Behavior*. New York: Holt, Rinehart and Winston.

Ekman, P. (1965) "Communication through nonverbal behavior: A source of information about an interpersonal relationship," in Tomkins, S. S. and Izard, C. E. (eds.), *Affect, Cognition and Personality*, New York: Springer, pp. 389–442.

Ekman. P. (1972) "Universals and cultural differences in facial expressions of emotion," in Cole, J. (ed.), *Nebrush Symposium on Motivarion*, 1971, Vol. 19. Lincoln: University of Nebraska Press, pp. 207–283.

Ekman, P. (1979) "About brows: Emotional and conversational signals," in Von Cranach, M., Foppa, K., Lepenies, W. and Ploog, D. (eds.), *Human Ethology: Claims and Limits of a New Discipline*, New York: Cambridge University Press, pp. 169–202.

Ekman, P. (1992) "An argument for basic emotions," *Cognition and Emotion*, vol. 6, no. 3–4, pp. 169–200.

Ekman, P. (2001) *Telling Lies: Clues to Deceit in the Marketplace, Marriage, and Politics*, 3d ed. New York: W.W. Norton.

Ekman, P. (2003) "Darwin, deception, and facial expression," *Annals New York Academy of Sciences*, vol. 1000, pp. 205–221.

Ekman, P. (2004) "Emotional and conversational nonverbal signals," in Larrazabal, M. and Miranda, L. (eds.), *Language, Knowledge, and Representation*, Dordrecht, Netherlands: Kluwer Academic Publishers, pp. 39–50.

Ekman, P. and Friesen, W. V. (1969) "The repertoire of nonverbal behavior: Categories, origins, usage, and coding," *Semiotica*, vol. 1, pp. 49–98.

Ekman, P. and Friesen, W. V. (1978) *Facial Action Coding System: A Technique for the Measurement of Facial Movement*. Palo Alto, CA: Consulting Psychologists Press.

Ekman, P. and Friesen, W. V. (1982) "Felt, false and miserable smiles," *Journal of Nonverbal Behavior*, vol. 6 no. 4, pp. 238–252.

Frank, M. G., Ekman, P. and Friesen, W. V. (1993) "Behavioral markers and recognizability of the smile of enjoyment," *Journal of Personality and Social Psychology*, vol. 64, pp. 83–93.

Half-Life 2 (2008) [Computer game], Valve Corporation (Developer), Redwood City, CA: EA Games.

Hampton, D. (2011) "Written all over your face," *Best Brain Possible*. Retrieved January 21, 2014, from http://www.thebestbrainpossible.com/tag/wallace-friesen.

Izard, C. E. (1971) *The Face of Emotion*. New York: Appleton Press.

Johnson, H. G., Ekman, P. and Friesen, W. V. (1990) "Voluntary facial action generates emotion-specific autonomic nervous system activity," *Psychopysiology*, vol. 27, pp. 363–384.

Kleinman, D. (director) (2013) *Galaxy Chaueffeur* [Animated advertisement]. London: Framestore.

Mori, M. (2012) "The uncanny valley," (MacDorman K. F. and Kageki, N. trans.), *IEEE Robotics and Automation*, vol. 19, no. 2, pp. 98–100. (Original work published 1970.)

Rozin, P., Haidt, J. and McCauley, C. R. (1993) "Disgust," in Lewis, M. and Haviland, J. (eds.), *Handbook of Emotions*, New York: Guilford Press, pp. 575–594.

Solid Angle (1999) *Arnold* (Illumination renderer). Madrid, Spain: Solid Angle S. L.

The Casting (2006) [Technical demonstration], Quantic Dream. (Developer), Paris: Quantic Dream.

Thompson, C. (2004) "The undead zone," *Slate*. Retrieved November 1, 2013, from http://www.slate.com/articles/technology/gaming/2004/06/the_undead_zone.html.

Tinwell, A., Grimshaw, M. and Williams, A. (2010) "Uncanny behaviour in survival horror games," *Journal of Gaming and Virtual Worlds*, vol. 2, no. 1, pp. 3–25.

Tinwell, A., Grimshaw, M., Williams, A. and Abdel Nabi, D. (2011) "Facial expression of emotion and perception of the Uncanny Valley in virtual characters," *Computers in Human Behavior*, vol. 27, no. 2, pp. 741–749.

Valve (2008) *Faceposer* [Facial animation software as part of source SDK video game engine]. Washington, DC: Valve Corporation.

CHAPTER 5

Applying Psychological Plausibility to the Uncanny Valley

T HE RESEARCH THAT I HAD CONDUCTED SO FAR on the uncanny and aberrant facial expression in human-like virtual characters raised intriguing questions as to the possible psychological cause of the Uncanny Valley. The results of an experiment described in the last chapter showed that the uncanny was found to be strongest in realistic, human-like male characters with a lack of movement in the upper face including the eyelids, eyebrows and forehead when compared with male characters with full facial animation or humans (Tinwell et al., 2011). This effect was particularly strong for characters communicating fear and surprise since without upper facial movement the salience of these emotions was reduced. Although this work provided guidance as to how the design of characters' facial expression may be improved to reduce uncanniness across different emotion types, the psychological drivers of the uncanny experience were still not fully explained. These results led me to question whether aberrant facial expression in a character may trigger possible psychological processes that underpin the existence of the Uncanny Valley. Furthermore, what specific personality traits may be perceived in an uncanny character with a lack of emotional expressivity in the upper face that may evoke a negative response in the viewer?

In this chapter, I first consider possible psychological explanations of the Uncanny Valley that other researchers have proposed, including a reminder of death (see, e.g., Mori, 1970/2012), perception of a threat (Kang, 2009), and an inability to empathize with a human-like agent (Misselhorn, 2009). Empathy has been defined as one of the unique qualities that constitute humanness and makes us human (Hogan, 1969). As such, being human is characterized by the ability to understand the cognitive and emotive processes of others and show tenderness, compassion and sympathy toward them, especially for the suffering or distressed. Given the importance of empathy in effective social interaction and previous works that suggest potential psychological explanations of uncanniness, I began to formulate a new hypothesis as an explanation for experience of the uncanny in human-like virtual characters. My hypothesis was based on a potential perception of psychopathic traits in a character and a lack of empathy in that character toward us (Tinwell, Abdel-Nabi and Charlton, 2013). To test this, in 2013 I conducted an experiment with psychologists Dr. Deborah Abdel-Nabi and Dr. John Charlton at the University of Bolton to explore which antisocial negative personality traits may be associated with uncanny human-like characters with aberrant facial expression and if a perception of psychopathy may be directly related to uncanniness. As well as considering the role of empathy in social interaction and specific physiognomic markers that signal a lack of empathy in an individual, this chapter provides an overview of the design, results and conclusions drawn from that empirical study. This previous study just focused on actors and videos of human-like video game characters within the age group of young adults. Therefore, in this chapter I also expand on how a perception of antisocial traits may affect characters intended to be perceived as empathetic of differing age groups in both animation and games. Last, I discuss how negative neurotic personality traits not associated with psychopathy may actually increase one's affinity toward the character, as one perceives them as more human and less of a controlled automaton.

5.1 PREVIOUS PSYCHOLOGICAL EXPLANATIONS OF THE UNCANNY VALLEY

In his original paper, Mori (1970/2012) theorized that we are equipped with the ability to experience the uncanny as an alert to possible proximal danger. In other words, it serves as a function to warn us of danger in close proximity rather than more distant sources of danger such as storms and hazardous weather.

> The sense of eeriness is probably a form of instinct that protects us from proximal, rather than distal, sources of danger. Proximal sources of danger include corpses, members of different species, and other entities we can closely approach. Distal sources of danger include windstorms and floods. (p. 100)

By using an example of a corpse as a dangerous object in the valley, Mori made clear associations of the uncanny with death; in this case uncanny near human-like characters may serve as a reminder of one's own death. This supposition is supported by more recent empirical work in that human-like agents can appear lifeless in their physical appearance. The results of a study by Karl F. MacDorman and Hiroshi Ishiguro in 2006 supported Mori's proposition and showed that certain features of human-like robots can represent an indication of death, thus inducing a reminder of death in the viewer. Similarly, male human-like virtual characters with a lack of facial expression may remind the viewer of a corpse and elicit the resulting feelings of dread associated with one's own inevitable mortality (Tinwell et al., 2011).

Building on Mori's (1970/2012) theory that uncanniness may be provoked by an immediate, close threat of danger, in 2009 Minsoo Kang purported that this perception of a threat was induced as human-like synthetic agents may go against one's predetermined world view. Rather than sitting comfortably in a category of what we understand to be human or a category for man-made mechanical objects, they instead lie between those categorical boundaries of what represents a human or machine. Hence, this state of cognitive dissonance alerts us to a possible threat, as we may not know how to interpret the object and how they will behave toward us. Kang supposed that the uncanny effect was in relation to how much control we perceive to have over this novel object. If we interpret that we are of a higher stature to an object and in control of a situation, then we may perceive there is little or no threat. Anthropomorphic characters such as Sonic the Hedgehog or WALL-E (Stanton, 2008) may be regarded as cute, comical and likeable as we perceive that we are of a higher status to that character and thus remain in control. However, there may be less certainty as to how much control we have when interacting with an android or human-like, virtual character. We may be less certain that we are more dominant than the human-like agent and that the agent does not intend to cause us potential harm. This may not only confuse us but also raise alarm, in that the object is now perceived as a threat, being more powerful than us and out of our control (Kang, 2009; see also Ramey, 2005). In this chapter, I provide evidence of antisocial personality traits associated with uncanny human-like, virtual characters that may make the viewer perceive that they are in a more vulnerable position when confronted with such a character.

In 2009, philosopher and professor Catrin Misselhorn presented a different perspective on uncanniness based on our inability to empathize with human-like, synthetic agents. Although we can anthropomorphize and empathize with nonhuman, inanimate objects, this affection diminishes as the human-likeness for a character increases. We may ascribe human characteristics and feel emotions toward a mechanical toy robot but may fail to maintain this empathy with a near human-like android (Misselhorn, 2009). As discussed in this chapter and beyond, I not only suggest that specific physiognomic markers may prevent a viewer from empathizing with a realistic, human-like character but also take Misselhorn's theory further: I propose that uncanniness occurs due to a perception of a lack of empathy *in a character toward us*. In the next section I explore what empathy stands for in humans and how we may perceive this quality in others.

5.2 EMPATHY AND HUMANITY

Professor of psychology Martin Hoffman (1987) described empathy as "an affective response more appropriate to someone else's situation than to one's own" (p. 48). In other words, empathy occurs as we recognize and respond to the emotional state of another individual. Having conducted a considerable literature review on the different forms of empathy, psychologist Robert James R. Blair (2005) defined cognitive empathy as theory of mind (p. 699). Cognitive empathy occurs when an individual can understand another's mental state and their likely internal thoughts (Blair, 2005). Emotional empathy relies on partially separable neural systems and can occur in response to "emotional displays of others" (p. 699) such as another's facial expression, speech and movements of their body. Other stimuli can also evoke an emotional response in an individual such as reading and responding to news stories; for example, "There were no survivors in the multiple-car accident" (Blair, 2005). One who is empathetic has the ability to appreciate (even share) the feelings of another, the cognitive ability to understand another's feelings and the social skills to show compassion toward another who is in distress (Blair, 2005; Decety and Jackson, 2004). In this way, empathy is a crucial component of effective and fluid social interaction (Caruso and Mayer, 1998; Davis, 1983;

Hogan, 1969; Mehrabian and Epstein, 1972; Thornton and Thornton, 1995). Therefore, perception of a lack of empathy in a character may suggest to the viewer that the character cannot experience the affective state of the viewer (experience emotional empathy), understand (cannot process cognitively) the thoughts (or feelings) of the viewer or demonstrate compassion toward them.

Antisocial personality disorder (ASPD), also referred to as psychopathy (Hare, 1980, 1991; Hart and Hare 1996; Herpertz et al., 2001), is a main psychiatric disorder associated with empathic dysfunction (Blair, 2005). Individuals diagnosed with psychopathy have distinct difficulties with emotional empathy and may show a lack of empathy toward people and animals (Blair, 2005). It has been found that emotions normally regulating one's behavior to avoid acting on impulse are lacking in those with personality disorders or psychopathy (Hare, 1970; Hart and Hare 1996; Herpertz et al., 2001; Lynam et al., 2011). This may make them more prone to violent outbursts as they resort to more violent and aggressive behavior to make themselves understood. As such, I proposed that the essence or cause of the Uncanny Valley lied in the perception of a lack of empathy in a human-like agent, which may suggest psychopathic tendencies in that character (Tinwell, 2014; Tinwell et al., 2013; Tinwell, Grimshaw and Abdel Nabi, 2014). Fear and panic may be instilled in the viewer if they perceive an atypical diminished degree of emotional responsiveness from a human-like, virtual character. With a perceived lack of empathy and without being able to predict the character's intended behavior from their nonverbal communication (NVC), uncanniness is exaggerated due to a proximal threat of potential dangerous and violent behavior, bordering on that of the psychopathological (ibid.). This theory also supported previous theories as to the psychological cause of the uncanny made by Mori (1970/2012), Kang (2009) and Misselhorn (2009). If interacting with an uncanny human-like agent, one may be aware of an immediate, possible danger as a perceived lack of empathy in a character works against what we would normally expect (i.e., our predetermined world view) in terms of "normal" human response and interaction. Furthermore, this unexpected response may prevent a viewer from empathizing with a realistic, human-like character, thus supporting Misselhorn's theory. To allow the reader to more fully appreciate the reasoning behind my theory, in the following section I review the major findings on psychopathic traits and then consider the similarities between these traits and those words used to describe uncanny human-like agents.

5.3 PERCEPTION OF ANTISOCIAL PERSONALITY TRAITS IN AN UNCANNY CHARACTER

Dr. Robert Hare, who has spent over 35 years researching psychopathy, not only introduced this classification to the domain of psychology but also asserts that it is a critical predictor of future risk and danger in those diagnosed with the condition (Hare, 1980, 1991). As a developmental order, psychopathy usually appears in childhood and may be confirmed by as early as eight years of age (Harpur and Hare, 1994). It has been found that this condition can continue throughout one's lifespan, and those with psychopathy represent 25% of those diagnosed with ASPD (Hart and Hare, 1996). Of particular importance to my investigation, a lack of empathy toward others is a fundamental diagnostic criterion of this condition (Blair, 2005; Hare, 1991). Such individuals typically show reduced guilt and compassion and increased behavioral disturbance toward others, involving criminal activity and violence (Blair, 2005; Hare, 1980, 1991; Harpur, Hare and Hakstian, 1989). With a blatant disregard and a lack of empathy or remorse for others, psychopaths commonly have an inability to forge meaningful attachments with others (Glenn, Kurzban and Raine, 2011). It is widely accepted that psychopathy is more common in males, especially the violent and impulsive aspects of psychopathy (see, e.g., Babiak and Hare, 2006; Buss, 2010; Cleckley, 1964; Glenn et al., 2011; Hamburger, Lilienfeld and Hogben, 1996; Hare, 1970; Hart and Hare, 1996; Lynam et al., 2011; Salekin, Rogers and Sewell, 1997). However, empirical studies have shown that female psychopaths may also demonstrate a lack of empathy and possess a manipulative and histrionic persona comparable to (or in some cases greater than) that of male psychopaths (Ekman, 1985; Hamburger et al., 1996; Salekin et al., 1997). While the behavioral aspects of psychopathy are male dominated, the deficits in social and interpersonal connection and communications are evident in *both* males and females (Ekman, 1985; Hamburger et al., 1996; Salekin et al., 1997).

Interestingly, I observed that many of the indices used in new scales to measure and describe *Shinwakan*, devised by those active in Uncanny Valley research, were comparable to those items used in scales to measure and describe psychopathy (Tinwell, 2014; Tinwell et al., 2013). Frustrated at the ambiguity concerning the translation of Mori's original neologism (*Shinwakan*) to describe uncanniness in 2009, Christoph Bartneck, associate professor at the Human Interface Technology Lab in New Zealand, and his colleagues developed a new scale named the Godspeed Indices

(p. 79) as a tool to describe and measure perceived human-likeness in androids. This scale included various alternative items to *Shinwakan* with a positive versus negative effect. Items such as dead–alive, dislike–like, unfriendly–friendly, unkind–kind and apathetic–responsive were established as semantic differential measures of uncanniness (Bartneck et al., 2009, p. 79). Building on this work, in 2010 Chin-Chang Ho and Karl F. MacDorman empirically tested a new set of indices that applied to both robots and animated realistic, human-like characters. This new set of items to measure the Uncanny Valley included the indices eeriness, attractiveness and humanness with synonyms used to describe each dimension (p. 1515).

The Elemental Psychopathy Assessment (EPA) scale was developed in 2011 by Dr. Donald Lynam, professor of psychology at Purdue University, and his colleagues as a self-report inventory for assessment of psychopathy. Based on the five-factor model of personality traits, I noticed that many of the EPA items bore similarities with items developed to measure and describe uncanniness in human-like synthetic agents. It seemed to me that there was a likely relationship between personality traits associated with psychopathy and descriptors used to measure the uncanny effect in human-like agents. As I disclosed in an earlier paper, specific traits to measure psychopathy such as anger and hostility, unconcern, coldness, callousness and distrust (Lynam et al., 2011, p. 113) are synonymous with those used in the uncanniness scales (Tinwell, 2014, p. 177). The items unfriendly and unkind from Bartneck et al.'s (2009) uncanny scale can be linked with anger and hostility (Tinwell, 2014). Perceived apathy or unresponsiveness (Bartneck et al., 2009, p. 79) in a human-like character maps directly onto unconcern as an ASPD trait (Tinwell, 2014). Uncanniness descriptors such as a dislike for a human-like character (Bartneck et al., 2009, p. 79) that is less agreeable and less predictable (Ho and MacDorman 2010, p. 1515) may be caused by a perceived distrust for that character. Psychopathic traits such as coldness and callousness are semantically associated with the uncanny characteristics eerie or spine-chilling in the Ho and MacDorman uncanniness scale (Tinwell, 2014). Given these strong thematic links between the EPA and scales used to measure the uncanny, I proposed that when we try to measure the uncanny in human-like characters we are actually assessing whether psychopathic traits are evident in a human-like character (Tinwell, 2014). In the next section I describe an experiment that I conducted with psychologists at the University of Bolton to test my theory that a perception of psychopathy is related to perception of uncanniness and how a lack of the startle response in the eye region and forehead to emotive situations was used as a specific visual facial marker of this condition.

5.4 LACK OF VISUAL STARTLE REFLEX AND PSYCHOPATHY

Normally, if we experience a sight or sound that frightens or surprises us, we automatically present a visual startle reflex that communicates our shock and heightened emotive state (Benning, Patrick and Iacono, 2005; Ekman, Friesen and Simons, 1985; Herpertz et al., 2001; Justus and Finn, 2007). In response to a sudden loud scream or seeing someone's shadow when we thought we were alone, spontaneously our eyes widen, pupils dilate and our eyebrows lift, causing creases in our forehead. This facial reflex occurs simultaneously with body gesture, such as leaning back or jumping out of a chair, as internally our blood pressure and heart rate increase (Ekman et al., 1985). However (and of particular importance to my theory), previous research has revealed that those diagnosed with psychopathy do not communicate this startle reflex (including wide eyes, raised brows and forehead) in response to frightening or shocking sights and sounds (Benning et al., 2005; Herpertz et al., 2001; Justus and Finn, 2007; Patrick, Bradley and Lang, 1993). In this way, they are less likely to experience a fear-induced reaction to aversive stimuli that those not diagnosed with psychopathy would typically show. Given that this inherent lack of the startle reflex in response to frightening or shocking stimuli is a known visual facial marker for those with psychopathic traits, I hypothetically linked this with the results of my earlier study on uncanniness in characters with aberrant facial expression. This earlier experiment revealed that the uncanny was strongest in a character when presenting the emotions fear and surprise without movement in the eyelids, eyebrows and forehead (Tinwell et al., 2011). Hence, the viewer may have perceived this lack of upper facial movement for fear and surprise as a physiognomic marker of a distinct lack of a startle response and therefore possible psychopathic tendencies in that character. Therefore, it may be this realization that contributed to experience of the uncanny, as it triggered an innate recognition of the potential danger and unpredictability associated with a psychopathic personality (Tinwell, 2014; Tinwell et al., 2013). This appraisal may not only have provoked uncanniness but also have provided a possible psychological explanation of the uncanny in human-like characters. However, this notion still has yet to be empirically tested.

Based on this, in 2013 I designed another experiment to see if I could replicate my earlier findings of 2011, this time using male and female characters to see if my theories applied to female characters too. Participant questions were included to assess possible psychological processes behind uncanniness by examining whether a lack of facial expression in the upper facial region (including the eyelids, eyebrows and forehead) may evoke perception of a particular type of negative personality traits in a character. Statistical models and analysis were then applied to the results to assess whether it was this appraisal that elicited an uncanny response in viewers.* A total of 96 female and 109 male participants (with an average age of 22.5 years) rated videos of a male and female actor, both 19 years old, and videos of a male protagonist character named Barney and a female protagonist named Alyx from the video game *Half-Life 2* (Valve, 2008). The virtual characters both had a near human-like appearance and age similar to that of the actors. We needed the actors to present a startled facial expression in the videos, so to achieve this we played the sound of a loud female scream and instructed the actors to present a startled response on hearing the scream. Importantly, on hearing this scream in each of the videos, it was likely that the participants would *expect a startled response* in a character. I used the Flex Animation tools within the facial animation software *FacePoser* (Valve, 2008) to copy the appropriate gender facial expressions of the male and female actors (in the human group) onto Barney and Alyx (in the fully animated group). Then, in addition to the human and fully animated virtual characters, videos were made for a partially animated group (named the lack group) in which movement in Barney and Alyx's eyelids, eyebrows and forehead was disabled.

The 205 participants watched these six videos, which consisted of two human videos (human group), two fully animated videos (full group) and two partially animated videos (lack group) in random order. For each video, participants then rated how much they agreed or not with 18 different statements used to describe each character. The statements included six items to measure perceived uncanniness, six items to measure negative personality traits associated with psychopathy and another six negative personality traits *not associated with psychopathy*. For uncanniness ratings, the items eerie, nonhuman-like, repulsive, unattractive, unlikeable and unresponsive were selected from scales devised to measure the

---

* For a full description of the methodology, design, procedure and results of this study please see Tinwell et al. (2013).

Uncanny Valley (see, e.g., Bartneck et al., 2009; Ho and MacDorman, 2010; Mori, 1970/2012). The personality traits angry, cold personality, dominant, uncaring, unconcerned and untrustworthy were taken from the EPA (by Lynam et al., 2011) to measure traits associated with psychopathy. To be certain that only negative traits associated with psychopathy evoked the uncanny, six items were also included from the NEO Neuroticism Facet Scale (Costa and McCrae, 1992). Thus, participants were also asked if they perceived other more negative personality traits in a character, not directly related to a psychopathic disposition, including anxiety, shame, depression, hopelessness, nervousness and self-consciousness.

Based on my previous experiments (see Tinwell et al., 2011 and Chapter 4 of this book), it was predicted that Barney and Alyx communicating a startled expression without upper facial movement (in the lack group) would be rated as most uncanny. Then, videos of these characters presenting fully animated startled expressions would be rated in second place, with the actor and actress being rated as least uncanny (Tinwell et al., 2013). However, to ascertain what was driving this perception of uncanniness it was expected that appraisal of personality traits associated with psychopathy would be a stronger predictor of the uncanny than character ratings of those other negative personality traits not associated with psychopathy. In the next section I describe the results from this experiment and the implications of this for human-like characters in video games and animation.

5.5 ABERRANT FACIAL EXPRESSION AND PERCEPTION OF PSYCHOPATHY

As was predicted, the human videos showing a startled expression were rated as least uncanny. The virtual characters, Barney and Alyx, were regarded as significantly more uncanny than the human videos when they communicated a fully animated startled expression in response to the scream sound. Yet the magnitude of this uncanniness increased further for Barney and Alyx when they presented a startled expression with a lack of upper facial movement (Tinwell et al., 2013). These results not only replicated the pattern that had been identified in the earlier study I conducted in 2011, as uncanniness increased from human to full to lack, but also showed that this pattern generalized to female characters with abnormal facial expression. Overall, in the human and full groups, male characters were rated as significantly more uncanny than female characters, although no gender difference was found for uncanniness ratings in the lack group. In other words, the actress and Alyx were regarded as less uncanny than the actor and Barney, yet Barney and Alyx were rated more closely for uncanniness when presented with only partial facial movement (Tinwell et al., 2013).

The purpose of this study was not just to provide design guidelines as how to control the uncanny in human-like virtual characters when presenting a startled expression but also to examine which particular negative personality traits may evoke perception of the uncanny. The results did confirm that the perception of psychopathic traits in a character was driving perception of the uncanny (Tinwell et al., 2013). Rather than other negative personality traits not associated with psychopathy, traits such as anger, coldness, dominance, callousness, an unconcerned attitude toward others or being of an untrustworthy nature were perceived in uncanny characters, especially those with a lack of expression in the upper face. On these grounds, we may be less tolerant of traits such as dominance or hostility in a human-like character and therefore be less comfortable in their presence. We may even feel the impulsion to get away from (or challenge) that character due to their more antisocial, callous demeanor. Hence, a new psychological explanation of the uncanny could be put forward in that a viewer may be alerted to possible psychological traits in an uncanny character due to that character's abnormal facial expression (Tinwell et al., 2013).

Conversely, participants were able to recognize nonpsychopathic negative traits such as shame, nervousness and being anxious in the human videos but not in Barney in Alyx. It seems that these more neurotic traits actually increased a character's humility (and reduced the uncanny), yet they were not effectively portrayed in the human-like virtual characters (Tinwell et al., 2013). We may recognize that someone is nervous or self-consciousness due to their less sociable and less confident behavior, but we may not perceive such traits as a potential threat toward us. We may even exude sympathy toward those who are despondent and have difficulty with social interaction due to their nervousness or self-consciousness. They are traits that arguably help to show one's vulnerability and, in that sense, make us appear even more humane! In this way, traits that suggest weakness and vulnerability rather than strength and power may actually increase one's acceptance and empathy for human-like virtual characters. We may ascribe the adage "They are only human" to the character, so this perceived weakness of being depressed, hopeless and/or self-conscious paradoxically increases our affinity toward them. However, in the case of this experiment, even in fully and expertly animated virtual characters, Barney and Alex's facial expressions were still less able to convey these more humane personality traits.

Out of the six characters, the female actress was rated as least uncanny (i.e., least eerie, nonhuman-like, repulsive, unattractive, unlikeable and unresponsive), but the male human-like virtual character Barney achieved the highest ratings for these uncanny characteristics when he was presented with a lack of upper facial expression. Given that (1) traits associated with psychopathic nature were found to be a significant strong predictor of Barney's uncanniness, (2) psychopathy is more common in males and (3) a lack of the startle response to shocking sights or sounds is a visual marker of psychopathy, this provided *prima facie* evidence that it was perception of psychopathy driving experience of the uncanny (Tinwell et al., 2013). As stated in the original paper, it is this acumen and perception of a threat that underlies experience of the uncanny:

> The viewer is repulsed or shocked by uncanny characters due to an appraisal and rapid response judgment that they are interacting with someone or something that does not display the typical reaction to shocking stimuli; someone with, possibly psychopathic-like traits who may present a potential threat to one's self-preservation. (p. 1623)

Again, this explanation echoed Mori's (1970/2012) interpretation that experience of the uncanny indicated a proximal threat and may serve as an avoidance strategy to prevent us being in potential danger. The possibility of being confronted with an angry, hostile, dominant, impulsive and unconcerned human-like virtual character alerts us to a nearby danger; hence, uncanniness occurs as an instinctive reaction to protect ourselves from this dangerous character. On hearing the female scream in the videos, a lack of the expected startle response in Barney and Alyx may have at least, even momentarily, increased perception of antisocial and unpredictable behavior in those characters. Hence, we avoid or reject those characters to protect ourselves from potential harm (Tinwell et al., 2013). In this regard, experience of the uncanny may bequest an integral, helpful survival function used in the real world (Tinwell et al., 2013).

In his original essay on the uncanny (and as I have previously discussed in Chapter 1 of this book), Freud (1919) associated the uncanny with thoughts, feelings or behaviors with which we may be familiar but have repressed to aid social interaction and our own well-being. "We shall find I think, that it fits in perfectly with our attempt at a solution, and can be traced back without exception to something familiar that has been repressed" (p. 247). In this instance, the uncanny may have occurred as a sinister revelation of repressed antisocial behavior that had been revealed in the human-like virtual character. Incongruence between the character's facial expression and their expected behavior (in this case a lack of a startle response to a scream sound) exposed traits that they may otherwise have been attempting to conceal. This leakage suggested a dangerously egocentric and sociopathic personality, one that goes against more positive social encounters. This also supports Kang's (2009) analogy of the psychological cause behind the uncanny, in that we may now perceive that we have less control over the more dominant character. We may feel intimidated by their Machiavellian persona and unpredictable behavior and seek to get away from that character (Tinwell et al., 2013). Now that perception of psychopathy has been empirically confirmed as a possible psychological cause of the uncanny, in the next section I address what the practical implications of this may be for human-like characters featured in games and animation. I also consider if this should be a primary concern for just male characters or if it extends to females and those characters of a younger and higher age group too.

5.6 THE EFFECT OF CHARACTER GENDER AND AGE ON UNCANNINESS

Those researching the condition psychopathy have found that there is an increased chance of extreme risk taking and aggression in male psychopaths (see, e.g., Benning et al., 2005; Hare, 1970; Herpertz et al., 2001; Justus and Finn, 2007; Patrick et al., 1993). This helped to explain why, overall, participants were less comfortable when viewing the male human-like virtual character Barney than his female equivalent, Alyx. Participants were aware not only of Barney's potentially cold and callous nature but also his hostile and aggressive behavior (Tinwell et al., 2013). A perception of these harmful traits in male protagonist characters intended by their designers to be perceived as empathic, such as Spielberg's (2011) human-like Tintin and Ethan Mars from the game *Heavy Rain* (Quantic Dream, 2010), may help elucidate why some viewers took a more negative reaction toward them. An apparent lack of a startled reaction on Tintin's face in response to the sound of loud gunfire and explosions when viewers would have anticipated Tintin to show a scared or shocked expression may have triggered this psychological driver of the uncanny. Moments when his face should have communicated the more ugly distortion of fear, anguish and shock are rather masked behind a more controlled, stoic representation of such negative emotions. Viewers questioned whether Tintin was really scared and shocked. His facial expressions lacked the intensity of fear and anguish that a viewer may expect given the dramatic and life-threatening situations that occur in the movie. The viewer may have been alarmed at Tintin's unmoved and cold demeanor, thus suspecting more antisocial traits such as a selfish, scheming and hostile personality. In the viewer's mind, there is a sense of incongruence and a feeling of something being not quite right about this character to the extent that he loses believability and credibility for being an empathetic character that we can engage with. In this case, it is a matter of context, and there is a mismatch between Tintin's facial expressions and what is happening to him in the film.

This also works against the human-like character Ethan Mars in *Heavy Rain* (Quantic Dream, 2010). His facial expressions do not communicate the emotions we may expect to see in response to traumatic events in the video game. Having lost his son, Jason Mars, in a busy shopping mall, Ethan Mars spots a red balloon that his son is holding floating above the heads of the crowds. As a possible sign of his son, Ethan follows this red balloon as it moves toward the mall exit and eventually sees Jason standing on the far side of a road outside the mall. Ethan immediately calls out to Jason, who starts to cross the road to rejoin his dad, but without first checking the traffic. Ethan looks to see a car fast approaching and in danger of hitting his son. Even though Ethan's gaze is focused on the car, there is a lack of a startle reflex on Ethan's face as the realization occurs that his son is in immediate danger and is likely to be hit by the oncoming vehicle. Ethan's mouth opens, but there is a lack of upper facial movement such as wide eyes, dilated pupils and arched eyebrows to visually communicate his fear and shock. We are hard-wired to detect such possible physiognomic markers of psychopathy, and the viewer may be subconsciously alerted to the fact that Ethan fails to show this expected startle reaction to this traumatic situation—hence, the viewer is now aware of possibly more sinister psychopathic traits in Ethan. One aspect that David Cage wanted to emphasize in *Heavy Rain* was the characters' emotional intelligence and to achieve "emotional immersion, to make you feel like you're with these characters, sharing what they feel" (Cage, as quoted in an interview with Neon Kelly, 2012). Furthermore, Cage was also keen to demonstrate a strong bond between Ethan and his family, especially in the early scenes of this game. Given that psychopaths often lack the ability to forge meaningful attachments with others, the viewer may have doubted whether Ethan had the ability to express empathy toward other characters and questioned whether he actually cared about Jason. This may have confused players as they struggled to understand this character's intentions and his defined role as a caring father, who was distraught about his son. In the case of *Heavy Rain*, I suggest that in the player's mind rather than, "here's Ethan," the phrase "here's Johnny"* may be more appropriate!

The results of the 2013 study by Tinwell and colleagues also showed that when uncanniness ratings were statistically compared for Barney and Alyx, when they presented a startled expression without upper facial movement there was no statistical significant difference between these two characters. Therefore, the female character Alyx was judged as just as strange and potentially threatening as the male character Barney when she was presented with aberrant facial expression. This suggested that stereotypical gender roles were of little importance during the rapid response judgment of potential psychopathic traits in an uncanny character. While female psychopaths are generally less common in society than male psychopaths, participants may have been wary of manipulative and controlling traits in *anyone* that displayed facial markers of psychopathy. Psychologist Joni E. Johnston (2013) stressed that female psychopaths may show a lack of empathetic concern and insincerity toward others above and beyond that of male psychopaths and that they may resort to violence if they cannot get their way through charismatic manipulation (see also studies by Ekman, 1985; Hamburger et al., 1996; Salekin et al., 1997).

> Male and female psychopaths are a lot alike in terms of their core personalities. They are self-centered, deceptive, shallow emotions, and lack of empathy. … female psychopaths are quite willing to resort to brutal violence to attain their needs when deceit, manipulation and charm either fail or are not available. (p. 1)

This is particularly relevant for human-like female virtual characters that are intended to fulfill the prototypical role of an empathetic spouse, friend, or mother in an animation or video game. Figure 5.1 shows an example of a female empathetic, human-like character typical of those used in games and animation.

---

* As spoken by the truly psychopathic character Jack Torrance in the film *The Shining* (1980).

FIGURE 5.1   An example of a female, empathetic human-like character typical of those used in games and animation by 3D artist Lance Wilkinson.

Milo's mother in the animation *Mars Needs Moms* (Wells, 2011) was criticized for showing a lack of emotional expressivity, to the extent that the viewer could not detect warmth or compassion from her toward her family. The virtual simulation of actress Joan Cusack failed to demonstrate a caring and nurturing disposition due mostly to her inadequate and bizarre facial expression. A lack of detail in Milo's mother's upper facial expression, especially in scenes when she was supposed to appear scared or startled, exacerbated the possibility of antisocial traits in this character. Rather than being capable of expressing empathy and love toward others and her family, the viewer may have been more suspicious of potentially manipulative and unpredictable traits in this character. The viewer may be left thinking that she does not care about her son or others and is concerned only about her own self-gain so they can no longer engage with her. Instead of being perceived as kind and likeable, she was criticized for being strange with a wax-work complexion, dead, emotionless eyes and with the emotional fidelity of a mannequin. In his article "Curse of the Mummy in *Mars Needs Moms*," Nicholas Schager (2011) drew connections between Milo's Mom and a chilling zombie character, more apt for a horror film. Schager described Cusack's computer-generated (CG) facial simulation as "unsettling" and stated that this character emanated a "creepy Madame Tussauds vibe" (p. 1). In an article for *USA Today*, writer Ryan Nakashima (2011) described feedback from parents who had taken their children to see *Mars Needs Moms*: "Doug McGoldrick, who took his two daughters to see the movie, said the faces of the main characters 'were just wrong.' Their foreheads were lifeless and plastic-looking, 'like they used way too much botox or something'" (p. 1). Another parent had explained to Nakashima that the human-like characters were "all annoying in their own way" (p. 1).

Milo's Mom was described as the living dead (Nakashima, 2011; Schager, 2011) as her behavior was perceived as frightening, annoying and disturbing. Due to this uncanny behavior, some viewers regarded her as more of a curse than a blessing (ibid.). In this instance, the viewer may have perceived psychopathic tendencies in Milo's mother that contradicted her defined role in this animation and that seemed more appropriate for an antipathetic, uncaring character. This may have helped ignite the criticism that this character (and others) in the film received because ultimately audiences were put off and dismayed at the sheer lack of emotional warmth that this family-oriented, feel-good animation was supposed to instill. The animation was intended to evoke sentimentality and emphasize the close bond between Milo and his Mom, yet the perceived lack of emotional empathy in these human-like characters left viewers cold and unmoved.

Furthermore, this appraisal may stretch beyond gender and be relevant to children and elderly human-like characters too. The actor and actress used in the study to test psychopathy as a predictor of uncanniness (see Tinwell et al., 2013) were each 19 years old, and the human-like virtual characters Barney and Alyx depicted young adults of a similar age. However, psychopathy has been determined as a lifelong condition that can present in individuals over the age of 60 and in children as early as the age of 8 (Harpur and Hare, 1994). In a recent article for *The New York Times*, journalist Jennifer Kahn (2012) wrote about Anne, whose 9-year-old son, Michael, had been diagnosed with psychopathy. Kahn also provided the perspective of Paul Frick, professor in psychology at the University of New Orleans, who has studied how to assess and treat psychopathic children. Anne described her son's behavior as extreme and was frustrated at the lack of emotional empathy in Michael toward others as well as his unpredictable outbursts: "Michael had developed an uncanny ability to switch from full-blown anger to moments of pure rationality or calculated charm—a facility that Anne describes as deeply unsettling. 'You never know when you're going to see a proper emotion,' she said" (p. 1). In support of this observation, Frick's work had identified the associated risks in such children who could show cruel acts toward others with a distinct lack of remorse. In Kahn's article, Frick described how psychopathic children can portray a blatant lack of care and empathy for others. "Callous-unemotional children are unrepentant. 'They don't care if someone is mad at them,' Frick says. 'They don't care if they hurt someone's feelings.' Like adult psychopaths, they can seem to lack humanity" (Frick, as quoted in Kahn, 2012, p. 1). A perception of psychopathic traits in either children or elderly virtual characters may be even more alarming for the viewer than in young adult characters, as characters of these age groups may typically be expected to be of a more naive and innocent (less threatening) nature. Children may be expected to have less worldly experience than adults and therefore be less scheming. In the case of the elderly, they may also be expected to be of a less calculating or sadistic persona due to potential symptoms of cognitive dementia and physical limitations that would typically make them more vulnerable to others. An example of an elderly male human-like virtual character is provided in Figure 5.2. Similarly, because of their comparative vulnerability, children or elderly members of the population may have greater sensitivity to uncanny characters as perception of psychopathic traits may pose a greater threat to those age groups. Based on this, it would be interesting to widen studies to include children and elderly human-like characters to see if psychopathy is a predictor of

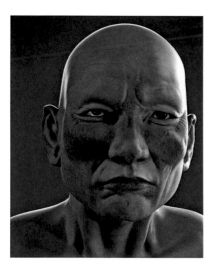

FIGURE 5.2   An example of a male, elderly human-like virtual character by 3D artist Lance Wilkinson.

uncanniness in these younger and older characters and to include participants of varying age in these future studies.

5.7 ANTISOCIAL TRAITS IN ANTIPATHETIC CHARACTERS

Character designers may consider this analogy of perception of psychopathy as a driver of the uncanny as a way to increase perception of antisocial
traits in antipathetic characters that are intended to have an uncaring and
scheming persona. Viewer perception that a character may be prone to
aggressive outbursts or capable of violent behavior with little remorse may
be beneficial for criminals featured in games and animation (Tinwell et al.,
2014). As stated in an earlier book chapter, "it may improve player engagement in a crime-thriller if those guilty characters may convey traits similar to those diagnosed with aspects of ASPD or psychopathy" (p. 335). By
reducing upper facial movement in monsters and villains when they may
be expected to show a startled response, designers may strategically apply
these tactics to evoke the uncanny and make evil characters appear even
more convincing. The player may benefit from criminals in crime-thriller
video games such as *LA Noire* (Team Bondi, 2011) that portray antisocial
traits, to improve not only the believability for the human-like criminal
characters but also player involvement and engagement with the story
(Tinwell et al., 2014). In game, the player can hold face-to-face interviews
with the *LA Noire* (Team Bondi, 2011) characters as a way to identify potential suspects or watch characters describe details of past events. In addition to what each character says, the player must rely upon observation of
a character's NVC to fathom that character's intentions and whether they
are guilty or innocent. In this way, aberrant facial expression in characters
with a realistic human-like appearance may help the player detect who is
a villain and which characters they may be more likely to trust (or not)
to help solve problems and win the game (Tinwell et al., 2014). The visual
startle reflex to shocking events may be omitted in antipathetic, criminal
characters such as Harlan Fontaine as a hint to the player that this character's behavior is suspicious and they are not to be believed (Tinwell et al.,
2014). Similarly, the depiction of a false smile in a character when they
should otherwise be happy and relaxed may guide the player to sense that
they are tense, scheming and not to be trusted. These facial visual cues
that suggest abnormal social interaction and ASPD in a character may
help the player understand which players would be capable committing
crimes. A hint of antisocial behavior to the extent of psychopathy in a
character (due to a lack of upper facial movement) may make the player

suspicious that this character is capable of violent, aggressive acts and killing others without regretting what they have done. Therefore, if a character's bizarre facial expression makes a player feel uncomfortable so that they are less trusting and accepting of a character, then uncanniness may be used by the player as a warning signal to detect a villain or killer in a game (Tinwell et al., 2014). This method of detection may make the player pay more attention to a character's facial expression and heighten player concentration and immersion in the game *LA Noire* (Team Bondi, 2011) and other similar crime-thriller games. This may help make such games more enjoyable and increase player satisfaction with that game.

> Solving a mystery in this way, by observing subtle nuances in NVC provocative of lies of omission, would no doubt be a more rewarding feat for the player, rather than having to rely solely on more blatant clues such as more obvious lies of commission. (Tinwell et al., 2014, p. 336)

However, this strategy would not be effective as a way to identify criminals in a game (or animation) if all characters, including those who were supposed to be empathetic, portrayed aberrant facial expression. As this is more often the case and most human-like characters, guilty or not, can unwittingly portray antisocial traits, designers should acquire a greater awareness of how we interpret NVC in the upper face, especially when showing a startled expression, before the uncanny can be used to help eliminate criminals. If the uncanny is to be used in this way, to the advantage of the player and designer, then the facial expression of empathetic characters that are not intended to mislead the player must be accurate and congruent with their in game state. If there is no clear difference in the facial expressions and perceived personas of 'good' versus 'bad' characters, then this more sophisticated element of player–character interaction may not be achieved and players may be left confused and frustrated with the game (Tinwell et al., 2014). Furthermore, improved graphical realism in subtle details such as pupil dilation, as well as the more obvious wide eyes and brow lift, may be necessary in characters before the uncanny may be controlled in this way.

The explanation provided here as to why viewers experience the uncanny is also prevalent to human-like virtual characters outside the domain of games and animation for entertainment purposes. For example, it is relevant to human-like relational agents used in e-learning applications and serious games for training and assessment purposes. Experience of the uncanny due to a perception of psychopathic traits in the relational agent that the participant is interacting with may have a detrimental effect on how that participant performs in an assessed task. The participant may be put off by the agent, thus preventing them from obtaining vital information from that agent and responding to them in an appropriate manner. The implications of this study stretch beyond that of human-like virtual characters but may also be applied to perception of the uncanny in android design and aberrant facial expression in humans. As such, the possible negative consequences of introducing androids designed to care for an aging generation need to be examined. Also, the potential risk of perception of more antisocial, psychopathic tendencies in those who have had Botox procedures to reduce movement (and wrinkles) in the forehead, between the eyebrows and the area around the eyes needs to be investigated.

Studies such as this to examine the possible psychological cause of the Uncanny Valley in human-like characters not only allow us to gain a better understanding of how we interact with human-like synthetic agents but also help reveal qualities that we accept as human. The fact that more neurotic (and less hostile) negative personality traits such as anxiety, nervousness, self-consciousness and shame were not perceived in Barney and Alyx offers support for Misselhorn's (2009) theory as a possible reason as to why we may longer empathize with characters with a more human-like appearance. Their inadequate facial expression fails to communicate such emotions, so we may therefore fail to express empathy toward them. Ironically, it is our perception of weaknesses in a character's personality that may make them appear less like an automated man-made object, less uncanny and more likeable. Here, and in the following chapters, I now take this theory further and suggest that we want human-like virtual characters to portray such emotions so that *we can perceive that they understand how we are feeling and thinking.* We want them to experience such negative personality traits so that they may be able to better relate to us and empathize with us if we experience more negative personality traits. As in real life, we do not want to be rejected or misunderstood, and we may be concerned that a human-like character that is incapable of feeling more vulnerable negative emotions may reject us for doing so! As a solution, we reject that character first to prevent him or her from making us feel any (more) uncomfortable. In the next chapter the concept of a perceived lack of empathy in a character toward us as the root cause of the uncanny in human-like agents is investigated further, examining aspects such as a lack of facial mimicry opportunity from the character to the recipient and how this may prevent us from forging an attachment with that character.

REFERENCES

Babiak, P. and Hare, R. D. (2006) *Snakes in Suits: When Psychopaths Go to Work*, New York: Harper Collins.

Bartneck, C., Kulic, D., Croft, E. and Zoghbi, S. (2009) "Measurement instruments for the anthropomorphism, animacy, likeability, perceived intelligence, and perceived safety of robots," *International Journal of Social Robotics*, vol. 1, no. 1, pp. 71–81.

Benning, S. D., Patrick, C. J. and Iacono, W. G. (2005) "Fearlessness and underarousal in psychopathy: Startle blink modulation and electrodermal reactivity in a young adult male community sample," *Psychophysiology*, vol. 42, no. 6, pp. 753–762.

Blair, R. J. R. (2005) "Responding to the emotions of others: Dissociating forms of empathy through the study of typical and psychiatric populations," *Consciousness and Cognition*, vol. 14, pp. 698–718.

Buss, D. M. (2010) "Human nature and individual differences: Evolution of human personality," in John, O. P., Robins, R. W. and Pervin, L. A. (eds.), *Handbook of Personality* (3rd ed.), *Theory and Research*, New York: Guildford Press, pp. 29–60.

Caruso, D. R. and Mayer, J. D. (1998) "A measure of emotional empathy for adolescents and adults," Unpublished manuscript.

Cleckley, H. (1964) *The Mask of Sanity* (4th ed.). St. Louis: Mosby.

Costa, P. T., Jr. and McCrae, R. R. (1992) *Revised NEO Personality Inventory (NEO-PI-R) and NEO Five-Factor Inventory (NEO-FFI) Professional Manual*. Odessa, FL: Psychological Assessment Resources.

Davis, M. H. (1983) "Measuring individual differences in empathy: Evidence for a multidimensional approach," *Journal of Personality and Social Psychology*, vol. 44, pp. 113–126.

Decety, J. and Jackson, P. L. (2004) "The functional architecture of human empathy," *Behavioral and Cognitive Neuroscience Reviews*, vol. 3, pp. 71–100.

Ekman, P. (1985) *Telling Lies: Clues to Deceit in the Marketplace, Politics, and Marriage*. New York: W.W. Norton.

Ekman, P., Friesen, V. W. and Simons, R. C. (1985) "Is the startle reaction an emotion?" *Journal of Personality and Social Psychology*, vol. 49, no. 5, pp. 1416–1426.

Freud, S. (1919) "The uncanny," in Strachey, J. (ed. and trans.), *The Standard Edition of the Complete Psychological Works of Sigmund Freud*, (1956–1974), London: Hogarth Press, vol. 17, pp. 217–256.

Glenn, A. L., Kurzban, R. and Raine, A. (2011) "Evolutionary theory and psychopathy," *Aggression and Violent Behavior*, vol. 16, no. 5, pp. 371–380.

Half-Life 2 (2008) [Computer game]. Valve Corporation (Developer), Redwood City, CA: EA Games.

Hamburger, M. E., Lilienfeld, S. O. and Hogben, M. (1996) "Psychopathy, gender, and gender roles: Implications for antisocial and histrionic personality disorders," *Journal of Personality Disorders*, vol. 10, no. 1, pp. 41–55.

Hare, R. D. (1970) *Psychopathy: Theory and Research*. New York: Wiley and Sons.

Hare, R. D. (1980) "A research scale for the assessment of psychopathy in criminal populations," *Personality and Individual Differences*, vol. 1, pp. 111–119.

Hare, R. D. (1991) *The Hare Psychopathy Checklist—Revised*. Toronto, Ontario: Multi-Health Systems.

Harpur, T. J. and Hare, R. D. (1994) "Assessment of psychopathy as a function of age," *Journal of Abnormal Psychology*, vol. 103, pp. 604–609.

Harpur, T. J., Hare, R. D. and Hakstian, A. R. (1989) "Two-factor conceptualization of psychopathy: Construct validity and assessment implications," *Psychological Assessment: A Journal of Consulting and Clinical Psychology*, vol. 1, pp. 6–17.

Hart, S. D. and Hare, R. D. (1996) "Psychopathy and risk assessment," *Current Opinion in Psychiatry*, vol. 9, no. 6, pp. 380–383.

Heavy Rain (2010) [Computer game]. Quantic Dream (Developer), Japan: Sony Computer Entertainment.

Herpertz, S. C., Werth, U., Lukas, G., Qunaibi, M., Schuerkens, A., Kunert, H. J., et al. (2001) "Emotion in criminal offenders with psychopathy and borderline personality disorder," *Archives of General Psychiatry*, vol. 58, no. 8, pp. 737–745.

Ho, C.-C. and MacDorman, K. F. (2010) "Revisiting the uncanny valley theory: Developing and validating an alternative to the godspeed indices," *Computers in Human Behavior*, vol. 26, no. 6, pp. 1508–1518.

Hogan, R. (1969). "Development of an empathy scale," *Journal of Consulting and Clinical Psychology*, vol. 33, pp. 307–316.

Hoffman, M. L. (1987) "The contribution of empathy to justice and moral judgment," in Eisenberg, N. and Strayer, J. (eds.), *Empathy and Its Development*, Cambridge: Cambridge University Press, pp. 47–80.

Johnston, J. E. (2013) "The psychopathic mother: Children growing up without empathy or love," *Psychology Today*. Retrieved February 21, 2014, from http://www.psychologytoday.com/blog/the-human equation/201307/the-psychopathic-mother.

Justus, A. N. and Finn, P. R. (2007) "Startle modulation in non-incarcerated men and women with psychopathic traits," *Personality and Individual Differences*, vol. 43, no. 8, pp. 2057–2071.

Kahn, J. (2012) "Can you call a 9-year-old a psychopath?" *New York Times Online*. Retrieved February 21, 2014, from http://www.nytimes.com/2012/05/13/magazine/can-you-call-a-9-year-old-a-psychopath.html?pagewanted=all&_r=0.

Kang, M. (2009) "The ambivalent power of the robot," *Antennae*, vol. 1, no. 9, pp. 47–58.

Kelly, N. (2012) "David Cage: 'I remember how scared we were,'" Videogamer. com. Retrieved February 27, 2014, from http://www.videogamer.com/ps3/heavy_rain/news/david_cage_i_remember_how_scared_we_were.html.

LA Noire (2011) [Computer game]. Team Bondi (Developer), New York: Rockstar Games.

Lynam, D. R., Gaughan, E. T., Miller, J. D., Miller, D. J., Mullins-Sweatt, S. and Widiger, T. A. (2011) "Assessing the basic traits associated with psychopathy: Development and validation of the elemental psychopathy assessment," *Psychological Assessment*, vol. 23, no. 1, pp. 108–124.

MacDorman, K. F. and Ishiguro, H. (2006) "The uncanny advantage of using androids in cognitive and social science research," *Interaction Studies*, vol. 7, no. 3, pp. 297–337.

Mehrabian, A. and Epstein, N. (1972) "A measure of emotional empathy," *Journal of Personality*, vol. 40, no. 4, pp. 525–543.

Misselhorn, C. (2009) "Empathy with inanimate objects and the Uncanny Valley," *Minds and Machines*, vol. 19, no. 3, pp. 345–359.

Mori, M. (2012) "The uncanny valley" (MacDorman, K. F. and Kageki, N., trans.), *IEEE Robotics and Automation*, vol. 19, no. 2, pp. 98–100. (Original work published in 1970.)

Nakashima, R. (2011) "Too real means too creepy in new Disney animation," *USA Today*. Retrieved April 27, 2014, from http://usatoday30.usatoday.com/tech/news/2011-04-04-creepy-animation_N.htm.

Patrick, C. J., Bradley, M. M. and Lang, P. J. (1993) "Emotion in the criminal psychopath: Startle reflex modulation," *Journal of Abnormal Psychology*, vol. 102, no. 1, pp. 82–92.

Ramey, C. (2005) "The Uncanny Valley of similarities concerning abortion, baldness, heaps of sand, and humanlike robots," paper presented at the Views of the Uncanny Valley: workshop, IEEE-RAS International Conference on Humanoid Robots, Tsukuba, Japan, December 5.

Salekin, R. T., Rogers, R. and Sewell, W. (1997) "Construct validity of psychopathy in female offender sample: A multitrait–multimethod evaluation," *Journal of Abnormal Psychology*, vol. 106, no. 4, pp. 576–585.

Schager, N. (2011) "Curse of the mummy in *Mars Needs Moms*," *New York Village Voice*. Retrieved February 27, 2014, from http://www.villagevoice.com/2011-03-09/film/curse-of-the-mummy-in-mars-needs-moms/full/.

Spielberg, S. (Producer/Director) (2011) *The Adventures of Tintin: The Secret of the Unicorn* [Motion picture]. Los Angeles, CA: Paramount Pictures.

Stanton, A. (2008) *WALL-E* [Motion picture]. Emeryville, CA: Pixar Animation Studios.

Thornton, S. and Thornton, D. (1995) "Facets of empathy," *Personality and Individual Differences*, vol. 19, no. 5, pp. 765–767.

Tinwell, A. (2014) "Applying psychological plausibility to the Uncanny Valley phenomenon," in Grimshaw, M. (ed.), *Oxford Handbook of Virtuality*, Oxford: Oxford, University Press, pp. 173–186.

Tinwell, A., Abdel Nabi, D. and Charlton, J. (2013) "Perception of psychopathy and the Uncanny Valley in virtual characters," *Computers in Human Behavior*, vol. 29, no. 4, pp. 1617–1625.

Tinwell, A., Grimshaw, M. and Abdel Nabi, D. (2014) "The Uncanny Valley and nonverbal communication in virtual characters," in Tanenbaum, J., Seif El-Nasr, M. and Nixon, M. (eds.), *Nonverbal Communication in Virtual Worlds*, Pittsburgh, PA: ETC Press, pp. 325–342.

Tinwell, A., Grimshaw, M., Williams, A. and Abdel Nabi, D. (2011) "Facial expression of emotion and perception of the Uncanny Valley in virtual characters," *Computers in Human Behavior*, vol. 27, no. 2, pp. 741–749.

Valve (2008) *Faceposer* [Facial animation software as part of source SDK video game engine]. Washington, DC: Valve Corporation.

Wells, S. (2011) *Mars Needs Moms* [Motion picture]. Burbank, CA: Walt Disney Pictures.

CHAPTER 6

The Mind's Mirror and the Uncanny

Very often we do manage to sense the inner states of others even though they try to hide them. We feel sadness behind a faked smile, or bad intentions behind seemingly generous actions. How do we do it? How do we manage to feel what is concealed?

KEYSERS (2011, p. 9)

THE RESULTS OF MY INVESTIGATION into why we may experience the uncanny in human-like virtual characters featured in games and animation had thus far revealed that we have a more negative response to human-like characters if we perceive a lack of empathy in that character toward others. We may perceive a lack of responsiveness in that character and an inability to show compassion toward others due to a lack of nonverbal communication (NVC) in their upper face, which we rely on to understand how that character is feeling and their likely behaviors. We may be wary of more negative, antisocial personality traits in a character due to their abnormal facial expression as a way to try to hide or conceal more negative emotions, such as insincerity, untrustworthiness or sadness behind a fake smile (Tinwell et al., 2011, Tinwell, Abdel-Nabi and Charlton, 2013). This may be to the extent that we perceive psychopathic tendencies in that character and a lack of concern for others as they fail to demonstrate a convincing startled response to fearful and shocking events (Tinwell, 2014; Tinwell et al., 2013). Recent findings in neuroscience and psychology offered support for my psychological evaluation of the uncanny in human-like virtual characters, and in this chapter I discuss how fundamental neurological and physical processes such as mirror neuron activity (MNA) and facial mimicry may be involved in our experience of the uncanny. With reference to work undertaken by neuroscientists such as Dr. Christian Keysers, professor in the social brain at the University Medical Center Groningen, I explore the role that the brain may play in uncanniness in human-like characters.

We instinctively mirror the actions of others as a way to understand their behavior. If vital information is missing in a human-like character such as a lack of facial expression and gesture, then this may result in cognitive conflict as we struggle to make sense of that character's actions. In this chapter, I also consider particular medical conditions that cause facial paralysis, thus preventing humans from creating facial expressions and mimicking those of others and the impact of this in social interaction— especially how the possible negative implications of a lack of mimicry opportunity in humans may relate to human–character interaction. This includes previous research in those with Moebius syndrome that helps to demonstrate how seemingly inconspicuous facial movements play a pivotal role in our daily lives. I provide an example of a human-like relational agent featured in a serious game designed for learning purposes where a lack of facial mimicry and emotional contagion in the virtual character may negate a pupil's ability to learn from and engage with that human-like character. Building on this, I put forward that uncanniness may not only occur in synthetic human-like agents but also in humans with abnormal facial expression, such as in those who have had facial cosmetic procedures such as Botox since it may prevent the opportunity for facial mimicry and shared emotional contagion.

6.1 MIRROR NEURON ACTIVITY

In the early twentieth century, the German philosopher Theodor Lipps (1851–1914) introduced the concept of an automatic motor mimicry imitation process, which allows us to understand the behavior and feelings of others. Lipps (1905) proposed that people instinctively mimic the posture, body movements, gestures, tone of speech and facial expressions displayed in others and that this mimicry response may elicit corresponding emotions in the observer. Since Lipp's initial theory of mimicry, neuroscientists and psychologists have continued to investigate the significance of how and the processes by which we interpret the behaviors and actions of others (see, e.g., Blake and Shiffrar, 2007; de Weid et al., 2006; Keysers, 2011). The actual discovery of mirror neuron activity (MNA) did much to explain this. It allows us to understand others' actions, intentions behind those actions and others' feelings (Iacoboni, 2009; Iacoboni et al., 2005; Keysers, 2011; Rizzolatti and Craighero, 2005). Mirror neuron cells were first discovered in the brains of macaque monkeys in the 1990s by a team of Italian researchers including the neuroscientists Giacomo Rizzolatti and Vittorio Gallese at the University of Parma (Di Pellegrino et al., 1992; Gallese et al., 1996; Rizzolatti et al., 1996; Rizzolatti and Craighero, 2005; Winerman, 2005). They found neurons in the premotor cortex that were activated not only when the monkeys grabbed an object, such as a peanut, but also when they observed others grab that same object (ibid.). This suggested that the monkeys were able to understand another's actions, such as picking up a peanut to eat, without conducting that action themselves (Di Pellegrino et al., 1992; Gallese et al., 1996; Rizzolatti et al., 1996). Since then, mirror neuron cells have been identified in humans and are located in motor regions of the brain including the inferior frontal gyrus and the ventral premotor cortex (Iacoboni et al., 2005). As found in experiments with the macaque monkeys (Di Pellegrino et al., 1992; Gallese et al., 1996; Rizzolatti et al., 1996; Rizzolatti and Craighero, 2005), these particular cells are activated in the brain not only when we carry out a particular action but also when we observe another performing that same action. For example, if we watch another person pick up a glass, then our own mirror neuron cells that would be used in picking up a glass are activated. If we see a person place a football on the ground, then it is likely that they are going to kick that football and the same neural circuits are fired in us as if we were about to kick the ball. The context in which an action takes place also helps us fathom why a person carried out a particular action and what they are likely to do next (Iacoboni et al., 2005; Rizzolatti and Craighero, 2005; Winerman, 2005). If a person picks up a glass in a café, then they are likely to drink from that glass. However, if they pick it up while standing at a kitchen sink, one may predict the glass will be placed in the sink to be cleaned. Therefore, MNA helps us to understand the why of an action observed in others and, importantly, to predict their future actions (Di Pellegrino et al., 1992; Iacoboni, 2008, 2009; Iacoboni and Dapretto, 2006; Iacoboni et al., 2005).

Many researchers have claimed that it is this MNA that makes us human: we may be biologically hard-wired to mimic others, and this underlies our ability to demonstrate emotional empathy toward another.

Therefore, our ability to read the minds of others and mirror their actions also affects us emotionally and allows us to connect with others. In his book *The Empathic Brain*, the neuroscientist Christian Keysers (2011) defined a causal link between MNA and our ability to empathize with others. He wanted to investigate how the emotions that other people experience can affect us too, in that they "spill over to us" (p. 8). We share others' positive and negative emotions as a subconscious, uncontrollable process and rarely have to think about it. "Most of the time when we share the emotions of others, you're not even consciously doing anything; it just happens to you" (pp. 9–10). This extends not only to family, friends and those close to us but to strangers as well. We too may grimace in pain if we see another struck in the face by a fast-paced ball. When we witness pain in another, our own neurological templates for experiencing pain are fired, allowing us to understand and sympathize with their thoughts, feelings and suffering (Gallese et al., 1996; Iacoboni, 2009; Keysers, 2009, 2011). As Keysers stated, "Sharing the affect of other individuals is deeply grounded in our human nature" (2011, p. 94). This may also help explain why beads of sweat form on our foreheads when watching a football player take a decisive goal kick in a cup title penalty shootout or why we may experience sadness when we see someone grieving a person's death.

To demonstrate this, Keysers (2011) used a clip from the James Bond film *Dr. No*. Despite different cultures and backgrounds, he observed that an audience would typically mimic the tension and discomfort that the protagonist, Bond, experiences in the scene. Asleep in bed, Bond is awakened by a tingling sensation on his face, onto which a large tarantula spider has crawled. The tarantula's sharp claws have left little indentions on Bond's skin as it grips to his face where droplets of Bond's sweat appear. It seems that Bond's heartbeat accelerates as he glances for objects that he may use to knock the spider off him. On observing people in the audience, Keysers could interpret from their facial expression and body movements that they too were experiencing the unpleasant and uncomfortable feelings that the Bond character is portraying.

> They all know that they are perfectly safe. And yet, just watching the scene, their heartbeats accelerate, they start sweating a little, their bodies become tense, and some even feel their arms start itching as if the spider's claws were brushing their own skin. (Keysers, 2011, p. 8)

Keysers's reference to MNA and film actors may be extended to human-like virtual characters in animation and games. In this way, we may share another's emotions both in real life and when we are watching human-like virtual characters in an animation or playing a video game. However, I reveal that it may be a lack of adequate emotional response in such characters that prevents us from mimicking and experiencing that character's emotions (or perceived lack thereof) and that this may contribute to uncanniness. Furthermore, we can still experience a real sense of danger or a threat from a virtual character (i.e., the uncanny) even though the uncanny character is being watched onscreen.

Keysers (2011) wanted to explore the role that the brain plays in automatically, intuitively sharing another's emotions so that they become part of us. To understand how we do this, Keysers suggested that a wider distribution of these mirror cells exists beyond the motor regions of the brain that help facilitate an exchange of sophisticated social emotions. A mirror neuron system has been identified in the insula, a part of the brain responsible for more negative emotions such as humiliation, embarrassment, guilt, shame, lust and pride. Keysers found that this part of the brain will mirror these negative emotions in the same way that it mirrors real physical pain (Keysers, 2011; Jabbi, Swart and Keysers, 2007; Wicker et al., 2003). When subjects observed a person's hand attempt to caress another yet saw someone else rudely push the hand away, mirror cells in the insula were fired to experience the pain of social rejection and the associated emotions such as humiliation (Keysers, 2011). As well as the role of the insula in experiencing the emotions that another is feeling, Keysers (and other researchers) have also identified the role that mirror neuron cells play in understanding another's emotive state from watching their facial expression. In the next section I discuss how this imitation process, which allows us to experience different emotions and sensations in response to another's behavior, also extends to facial expression and how a lack of this motor mimicry may alert us to the uncanny in a human-like virtual character.

6.2 FACIAL MIMICRY AND EMOTIONAL CONTAGION

In day-to-day life, we typically engage in conversations or brief emotional exchanges with others. This may vary from a passing smile and nod to the receptionist on your way in to the office to a lengthy discussion about a work project with your boss. In either case, we may use facial expression to decipher if someone is in a good mood or if that favor that we were going to ask of them may be best left until another time. A person may smile as we approach them, but their lowered brow suggests that they may not be happy and would prefer to be left alone. But how do we ascertain a person's mood and respond appropriately to them? Even in brief social exchanges, our mirror system is at work on a subconscious, cognitive and physiological level to ensure smooth and effective social interaction.

During face-to-face social interactions, the results of previous studies show that MNA takes place whereby we mimic the facial expression of the person we are communicating with as a way to establish mutuality with that person (Blakeslee, 2006; Gallese et al., 1996; Giles, Coupland and Coupland, 1991; Keysers, 2011). When we communicate with others, we involuntarily mimic their facial expressions to help gain rapport with them. MNA automatically stimulates the same neural circuits and thus elicits the same (or similar) facial expressions as the person with whom we are communicating. Many researchers argue that it is this facial muscular activity that allows us to experience the same emotions as another and refer to this process as emotional contagion (see, e.g., Hatfield, Cacioppo and Rapson, 1993; Hess and Blairy, 2001; Hsee, Hatfield and Chemtob, 1992; Keysers 2011; Laird et al., 1994). In this way, we can determine that someone is happy to talk to us and that they are genuinely interested in our conversation, or we can take the hint that they are distracted and impatient and may prefer not to engage in conversation with us. Activity in the facial nerves and muscles may involuntarily stimulate concomitant emotions in the receiver, and this serves as a way to understand the cognitive and emotive state of the sender—that is, to *empathize* with another. In this way, scientists propose that facial mimicry is a three-stage process. It starts with the movement of facial muscles due to mimicry of expressions that conducts feedback to neural facial receptors involved in facial movement that then elicits appropriate emotions in the observer (Hatfield et al., 1993; Wild et al., 2003). For example, if we witness another weeping, our own face may distort to a sad expression as the corners of our mouth automatically droop. Thus, our own brain templates for experiencing sadness are activated, and we may feel sadness or be moved to cry ourselves. We intuitively understand their sad emotional state and can respond appropriately. It may be inappropriate to smile and laugh or to show no response at all, when confronted with a person who we understand to be upset and worried; instead we may offer sympathy toward them. Conversely, if a person is beaming with happiness on having passed an exam, the corners of our mouths may lift, and we too may experience a sense of happiness just from their smile. It may be seen as uncaring or unsociable if we do not share their happiness and offer little or a negative response to their happy state.

The theory that a smile is contagious led Barbara Wild, professor of neurology and psychiatry and psychotherapy, to investigate why smiles in others have the power to lift our own mood and how we may share humor. In 2003 Wild and colleagues conducted an experiment where participants viewed photographs of humans presenting happy, sad and neutral expressions. Two techniques were used to measure participant response: facial electromyography (EMG), a technique that measures the changes in the electrical activity of a person's facial muscles as they contract; and functional magnetic resonance imaging (fMRI), a technique that allows scientists to scan and measure a person's brain activity. In this case, scans were taken of the parietal cortex area of the brain thought to contain mirror neurons. The EMG readings showed significantly greater and *faster* upward movement of the mouth corners when viewing a smiling face than when viewing neutral or sad expressions. Intriguingly, there was an increased downward movement of the corners of the mouth and at greater speed when viewing sad expressions, compared with viewing neutral or happy expressions. The fMRI data revealed that in addition to high-order visual areas (the fusiform gyrus) and regions of the brain involved in emotional processing (including the putamen), the premotor areas associated with facial mimicry were activated when viewing happy faces. These findings intimate that humans have the ability to spread good humor or positive feelings via facial expression and that facial mimicry contributes to this process (Wild et al., 2003). Therefore, when someone smiles, tells a joke, or falls about in hysterical laughter, it is a combination of these involuntary facial movements to mimic their happy face, and the activation of mirror neuron cells to simulate happiness in us, that allows us to share and reciprocate a happy, humorous response as it if was contagious.

Similarly, experiments to investigate the role of facial mimicry and anger revealed that when participants were shown photographs of humans with angry facial expressions, facial EMG measurements showed an increased muscular activity in participants' upper facial region (Lundqvist, 1995). This included movement of the corrugator supercilii muscle to push the eyebrows down and inward to create a frown expression, as if the participants were attempting to portray an angry facial expression themselves (Lundqvist, 1995). In *The Empathic Brain* (2011), Keysers also described one of his experiments in which when participants observed others presenting the emotion disgust, it ignited the same neural pathways in the participants as if they were experiencing disgust firsthand. While in an fMRI scanner, participants watched a film clip in which an actor sniffed the contents of a foul-smelling glass. The actor moved his head away quickly from the glass, at the same time wrinkling his nose and turning the corners of his mouth downward to show disgust. Presented with this disgusted expression, participants may have understood that it was an unpleasant, bad smell that caused the actor to feel disgusted. The fMRI scans revealed that the same mirror neuron cells in the visual and premotor area of the brain that would be used cognitively when making a facial expression of disgust were fired in participants (Keysers, 2011). Importantly, mirror cells were also activated in the participants' anterior insula, which is necessary for experience of disgust and experience of the sensations associated with smelling a bad or disgusting smell. As Keysers stated, this helped to explain how facial mimicry caused an observer not only to make the same (or similar) facial expressions as those that they are watching but also to actually feel that emotion as part of emotional contagion.

> We had unraveled what in the brain causes what the psychologists call facial mimicry, vicarious activity in the premotor cortex as if doing the same facial expressions, and what they call emotional contagion, vicarious activity in the insula as if feeling the same emotions. (p. 98)

For facial mimicry to take place, we rely heavily on NVC in another's face. As discussed in Chapter 4 of this book, NVC primarily takes place in the upper face in the eyes, eyelids, eyebrows and forehead as the mouth and lower facial region may be constrained by speech. In this way, we subconsciously analyze and interpret those voluntary (and involuntary) facial actions of others, from intense frowns to subtler, fleeting movements of the brows. Keysers (2011) stressed that MNA plays an integral part in our ability to empathize and connect with others and that "a dysfunction of these cells can lead to an 'emotional disconnection' with others" (p. 11). Given the importance of facial mimicry to effectively understand the emotional state of others and extensive empirical evidence to support the theory that facial mimicry occurs and aids emotional contagion, in the next section I consider how humans who lack an ability to mimic others' facial expressions may cope with and adapt to this potential socially limiting factor. Building on this, I then relate a lack of mimicry opportunity and perceived emotional disconnection in humans to perceived uncanniness in human-like virtual characters.

6.3 A LACK OF FACIAL MIMICRY IN HUMANS

In 2010, Dr. Kathleen Bogart, a social worker and assistant professor of psychology at Oregon State University, was interviewed by *New York Times* journalist Benedict Carey and provided insight as to how a breakdown in communication can occur without an opportunity for facial mimicry. Bogart has had Moebius syndrome from birth, a rare congenital condition that causes paralysis of the face. This condition causes those affected by it to be unable to smile, raise their eyebrows or frown, so they are unable to present different facial expressions (Bogart and Matsumoto, 2010a). At the moment there are approximately 2000 known cases of this condition, two-fifths of whom live in America (Bogart and Matsumoto, 2010b). As well as experience of living with the condition, Bogart has conducted empirical studies to explore the implications of this condition in social interaction. She described a situation where a perceived breakdown in communication, or emotional disconnect, had occurred between herself and others who were in a distressed and anxious state and needed to see an empathic response from her. When assigned on a mission in 2005 to help refugees of Hurricane Katrina in the city of Baton Rouge, Louisiana, Bogart realized a sense of frustration and disappointment in those whom she was trying to help. Left with just a handful of personal possessions and forced to evacuate their homes, the refugees sought more than just the fundamental aspects of food and shelter and also needed a sympathetic and caring attitude from those social workers assigned to help them (Carey, 2010). As well as providing practical advice, they needed Bogart to recognize their trauma and feel their distress and loss. Yet, on meeting Bogart, the refugees could sense that she seemed emotionally cut off from them. Without the opportunity for facial mimicry in Bogart's face to reflect their pain, sorrow and loss, the refugees were uncertain as to whether Bogart really understood what they had been through. She could sense their torment as they searched her blank, motionless face for a sign of reciprocity. Bogart explained, "I could see the breakdown in the emotional connection between us, could see it happening and there was nothing I could do" (Carey, 2010). Bogart felt helpless as she watched the sad expressions of the refugees but was unable to return a sad expression to them as a way to show that she truly understood their grief: "I wasn't able to return it. I tried to do so with words and tone of voice, but it was no use. Stripped of the facial expression, the emotion just dies there, unshared. It just dies" (Carey, 2010). Without the ability to mimic the facial expressions of another, Bogart intimated that the emotive value of their communication was lost to the extent that it was dead and nonexistent.

The scenario that occurred with the refugees in Baton Rouge resonated with Bogart to the extent that she wanted to explore and decipher the importance of facial mimicry in affective social communication. Now, as part of her research at Oregon State University, Bogart is investigating the possible negative psychological and social implications of facial paralysis and facial movement disorders during social interaction. Importantly, the findings from Bogart's work so far suggest that those with Moebius syndrome find other ways to adapt and understand other's emotions. In 2010, Bogart conducted an experiment with psychologist David Matsumoto in which 37 people with Moebius syndrome observed 42 online photographs of different facial expression of emotion, such as anger, happiness and sadness (Bogart and Matsumoto, 2010b). The results showed that those with the condition could still recognize expressions in the photographed faces at the same recognition rate as those without the condition and also that no relationship was identified between the severity of the participant's facial paralysis with expression recognition. Bogart and Matsumoto concluded that without the use of the facial muscles to mimic other's facial expressions such as anger, happiness and sadness, other senses become sharper so that they may still sense an emotional charge from another person. Alternative systems may develop in the brain as a way to recognize facial expressions, and those with facial paralysis learn how to use these systems. During conversation, they may also concentrate more carefully on other factors such as the sender's eye contact, body movements, posture and tone of voice.

Bogart and Matsumoto's (2010b) theory about using alternative ways to facial mimicry to understand another's emotional state was based on the results of previous studies by other researchers on facial mimicry in those with autism. The results from a study by Daniel McIntosh and colleagues (2006) showed that while people with autism do not use facial mimicry for emotion recognition, given sufficient time they can still correctly identify photographs of different facial expressions. Instead, they rely on areas of the brain involved with voluntary processing and judgment making rather than a speedier automatic, social-emotion process. Building on this work,

Bogart and Matsumoto suggested that those who have lived their entire lives with facial paralysis and without the ability to mimic other's facial expressions have developed alternative neural pathways as a compensatory mechanism for emotion recognition, similar to those with autism. Rather than MNA occurring as an automatic, subconscious process, those with Moebius syndrome may use regions of the brain associated with *intentional* attention processes as a way to correctly identify other's emotions. Bogart and Matsumoto also stress that these compensatory mechanisms have developed over time since birth and that those who encounter a new or temporary inability to mimic the facial expression of others may struggle to develop such sophisticated alternative judgment mechanisms as those who have had to rely on such tactics since infancy. However, given that this process is intentional rather than automatic, they may take longer in identifying people's expressions than those without the condition. Such a delay in responding to others' emotions may be perceived as odd or strange and may contribute to difficulty in social communication. Furthermore, this may serve as a potential limitation in social interaction when it is important to respond promptly to another's behavior, for example, if another is angry and hostile and may act aggressively.

I put forward that the findings from studies into emotion recognition and those with Moebius syndrome are relevant to human-like virtual characters and other human-like synthetic agents designed for human interaction such as androids. In their research, Bogart and Matsumoto (2010a) refer to descriptions made by psychotherapist Matthew Joffe, who has Moebius syndrome, and to other literature to substantiate that the quality of social interaction may be reduced when interacting with a person with Moebius syndrome. If there is a lack of facial mimicry, then the interactant not only may be confused and risk misinterpretation of what the person with the syndrome is trying to communicate to them but also may doubt that they are being fully understood. It was explained that Joffe was wary of people not understanding him and his nonverbal cues, to the extent that Joffe could find social interaction as "a tiring, effortful exercise in impression management and minimizing understandings" (Bogart and Matsumoto, 2010a, p. 140). People with Moebius syndrome may also be concerned if others cannot understand when they are being humorous or sarcastic as they do not want to risk aggravating people. Even when those with Moebius syndrome are not attempting to joke or be sarcastic, a pleasant greeting such as, "it's good to see you," may be perceived as rude or sarcastic without the appropriate facial cues to show that it was intended as a warm and friendly gesture (Bogart and Matsumoto, 2010a). Based on this insight and research, Bogart and Matsumoto stated, "Mirroring and nonverbal synchrony are important for developing rapport, empathy, and emotional convergence between interacting individuals" (p. 140). Without reciprocal facial mimicry from the person with the syndrome, the inter-actant may become frustrated and discouraged by this confusing social interaction (Bogart and Matsumoto, 2010a).

Similar to encounters with those with Moebius syndrome, even if a character is programmed to detect a sender's emotional state from aspects such as their facial expression, tone of voice, words spoken or typed, and/or body language, a character may be perceived as strange or odd if they fail to show congruent facial mimicry. Rather than this feedback from the sender increasing the level of connection with the human-like character, as with the example of Bogart and the refugee, the level of con-nection diminishes to the point that it is broken or dead, as the sender is not receiving the facial feedback that they require to comfortably con-tinue with the interaction. Even if human-like virtual characters are pro-grammed to be able to respond to the sender with appropriate mimicry response, then this feedback should be synchronized simultaneously with the sender's actions as if it was a subconscious, effortless, automatic pro-cess. Any foreseen delay (or lack of response) may confuse and irritate the sender and raise concern that they are not being understood. They may perceive that the human-like character that they are attempting to interact with (as they would a human), has maladaptive, dysfunctional social skills and lacks the capacity to understand and connect with them.

6.4 FACIAL MIMICRY IN RELATIONAL HUMAN-LIKE CHARACTERS

In addition to animation and games, where the viewer may expect to see appropriate mimicry response from one character to another dur-ing social interaction, this factor is particularly prevalent to human-like relational characters used in serious games for learning purposes. A lack of facial mimicry response may have contributed to the feedback provided by 13 male and female 12- and 13-year-old students about a 3D, human-like relational character designed to represent the world-renowned author Hans Christian Andersen (HCA). Working with her team, Andrea Corradini's intention was to develop an educational environment using a game-like system interface to allow for informative communication and natural human–computer interaction between humans and embodied historical characters (Corradini, Mehta and Robering, 2009). The movements and speech from an HCA actor were used as a guide to recreate a 3D human-like HCA character who resided in his virtual study, created with a high level of graphical realism in keeping with HCA's realistic, human-like appearance. Students could explore HCA's study from a first-person perspective and ask him questions within his knowledge domains about his own life and literature works via speech, gesture (e.g., pointing) and the keyboard. It was intended that interaction with the HCA character and e-learning environment would improve students' literature skills and general knowledge and that this knowledge would be learned and retained to a greater extent than if they had used more traditional learning methods such as text books and teacher presentations (Corradini et al., 2009).

Corradini et al. (2009) stated that, in addition to synthesized speech and gesture, the HCA character could present different facial expressions and was designed to react emotionally to those interacting with him.

> HCA reacts emotionally to the user input by displaying emotions and by employing a meaningful combination of synchronized verbal and non-verbal behaviors. He can get angry or sad because of what the user says, or he gets happy if the user … likes to talk about his fairy tales. (p. 354)

Even though HCA may have offered some facial mimicry in response to the students' questions, such as smiling if he heard a title of one of his fairy tales mentioned by the student or getting angry or sad if he suspected criticism of his work, was this sufficient to convince the students that the virtual character fully understood what they were saying? Would his response always be appropriate to students' emotive states and the context of what they had said? When Corradini and her team interviewed the students about their interaction with HCA and the learning environment, some student feedback suggested that this may not have always been the case. Interestingly, even though the HCA character had been designed with the intention to be perceived as a friendly, approachable and empathetic character, some of the students commented that he was rude and impolite. Their interaction with HCA was "… good enough but he (HCA) is not the most polite person around" (Corradini et al., 2009, p. 366). Furthermore, some students also explained that, given what was going on during their interaction with HCA, he did not respond in a way that they expected him to. This was to the extent that some students presumed that HCA did not acknowledge their presence and could not understand them:

> Some reports point out a few cases where the character did not act upon the wish and expectation of the player. They felt that their interaction was "… frustrating when he did not answer my questions …" and HCA "… didn't understand everything". (p. 366)

I put forward that a lack of facial mimicry in HCA in response to the students, which the students may have expected to see, reduced the students' ability to effectively interact with this character. A lack of facial mimicry in HCA may have also contributed to the fact that some students thought that HCA was impolite and failed to understand them. Corradini et al. (2009) noted in their paper that previous research into children's speech and communication had revealed that, when compared with adults, children's voices have a much greater range of prosody and acoustic characteristics (Darves and Oviatt, 2002). In other words, they may be more expressive and use more emotive tones over a greater range of pitch than adults. Children may be less fluent in their speech, with more pauses or hesitation (Oviatt, Darves and Coulston, 2004), especially in children of a more shy and introvert nature that may be less confident and less reluctant to engage with a virtual character (Darves and Oviatt, 2004). Overall, children may be less polite in social interaction and communicate their frustration more readily via speech and their behavior (Bell and Gustafson, 2003). Corradini et al. stressed that a potential limitation of HCA's communication lay within the restrictions posed by the natural language understanding (NLU) module within the system: "The proposed NLU is not capable of capturing fine distinctions and subtleties of language since it cannot produce a detailed semantic representation of the input utterance" (Corradini et al., 2009, p. 360). The NLU module could not process an exhaustive range of the words used, their prosody and emotive intonation within the incoming speech (Corradini et al., 2009).

Based on the previous discussion, would HCA have been able to respond appropriately if the students hesitated or paused between words or sentences at any point during conversation? Even if a student was silent but presented a thoughtful expression while pondering what to say next, would HCA have maintained eye contact with the student and raised his brows to signal that he was interested in what the student was trying to say? With children's tendency for more disfluent speech, if there was a delay in conversation from the student did HCA use NVC to effectively portray that he was still engaged with and listening to that student? If the student was struggling to find an appropriate word, raised brows and a smile may have shown encouragement from HCA, thus reducing anxiety for the student. A reciprocal response to demonstrate empathy and understanding from HCA toward the student may be even more important to encourage shy students who may lack confidence when interacting with relational, human-like characters. If HCA just presented a blank expression while waiting for recognition of a student's words before he responded to them, then the student may have perceived that HCA was expressing sarcasm or being rude. In the student's mind, not only had HCA lost interest in them, but also he was mocking them as they struggled to find the appropriate words and questions to ask him. He did not understand that they were doing their best to ask HCA questions, and he seemed disinterested in them. If a student perceived that HCA was bored with them due to a lack of facial mimicry response from him to communicate interest and encouragement, then he may have been perceived as impatient and unfriendly, qualities that work against the role and purpose of this human-like relational character.

Aware that "depressed or anxious children cannot assimilate new knowledge and learn as effectively" (Corradini et al., 2009, p. 367), Corradini and her team stressed the importance of appropriate display of emotion in a virtual character to improve "the effectiveness of computer-based learning environments populated with virtual agents" (pp. 367–368). However, possible limitations in HCA's ability to mimic the emotive state of the students and to empathize with them may have made the students more anxious, and reduced their ability to learn from him, thus impeding the learning process. Some students may have grown increasingly frustrated and anxious at not being able to find the right words quickly and may have needed to see empathy from HCA to show that he understood that they were anxious and were trying hard. For example, a student may have expected to see a concerned expression from HCA in response to their unhappy or worried expression as the student became more anxious. But a lack of a sympathetic and helpful response from HCA may have just frustrated or upset the students further.

Children can use a more animated range of vocal tones and facial expression during social interaction than adults, and if HCA did not communicate a similar emotive response to them then his behavior may have been perceived as odd. For example, if a student greeted HCA with a jovial and loud "Hi there!" while beaming a large smile but he greeted them back with a much less intense smile and a more controlled, monotone voice, then the student may have perceived that HCA had a colder, stricter personality and may not have understood how happy and excited they were to see him. If, while exploring HCA's study, a student discovered a secret book to help solve a task that made them gasp in delight but HCA did not share their excitement as the student may have expected him to and he appeared unmoved, then the student may have perceived that he was not aware of them or their actions.

With regards to the character graphics and believability, Corradini et al. (2009) stated that the increased realism increased how entertaining HCA (and the learning environment) were perceived to be, yet there were noticeable limitations in HCA's behavior that did not match his human-like appearance. "Children appreciated the life-like animations and graphical appearance of the character judging that 'The good graphics also makes it (the system) entertaining …' However, the repertoire of HCA's actions was sometimes perceived as rather limited" (p. 366). Other factors such as speech recognition, body movements and limitations in HCA's knowledge domain may also have contributed to this feedback in how limited HCA was perceived to be. Yet I propose that a perception of a lack of facial mimicry and reciprocal response to show that HCA shared and understood the students' emotions and thoughts may have reduced the overall believability for this character. This not only may have put the students off interacting with HCA but also may have raised anxiety in students and had a negative impact on their ability to learn from him and the e-learning environment.

To ensure smooth interaction with HCA (or other human-like relational characters) as well as to keep track of what has been said, the system interface has to plan the next response and simultaneously present and update its emotional state (Corradini et al., 2009). The modules designed to carry out these multiple tasks need to run in synchrony as any delay in (or lack of) mimicry response may be perceived as effortful and unnatural, putting the user off. In the same way as in real life, when the quality of social interaction may be reduced if there is a perceived lack of facial mimicry in those with Moebius syndrome (Bogart and Matsumoto, 2010a), the interactant may perceive that the human-like relational character lacks the ability to understand them. As well as feeling misunderstood, they may regard the character as insincere and rude due to their limited or inappropriate response to them. Those creating human-like relational characters have acknowledged that it is a great challenge for the research community and developers to deliver appropriate, engaging multimodal behavior in a human-like character that meets every possible situation in social interaction, especially to prevent interruption of the flow of interaction and misinterpretation between the character and human (Allen et al., 2001; Busine, Abrilian and Martin, 2004; Corradini et al., 2009). An increase in detail and accuracy of reciprocal NVC in the upper face, if the lower face is constrained by speech, may improve a student's engagement and interaction with a human-like relational character and, in this way, may help their degree of learning from the character. If students become frustrated at not being acknowledged and understood, due to (among other things) a lack of facial mimicry in a character, then this may have a detrimental impact on their ability and willingness to learn from a human-like relational character. Unless a seamless and contextual mimicry response can be achieved with human-like relational characters (such as HCA) then I suggest it may be advisable not to include characters with a highly human-like appearance in applications designed for learning purposes. While increased levels of character realism may create an initial wow factor and impress students, can student expectations then be fulfilled and maintained? Feedback from students about HCA may have been more positive had the level in realism in his appearance been reduced to match his more limited behavior. Therefore, a more simplified human-like appearance may have reduced expectations in the students as to how HCA should respond to them. A reduced human-like and more stylized appearance (less like Milo's Mom and more like Mr. Incredible) may be beneficial for such characters to aid student interaction and learning. However, more work is required to assess the effect of pursuing increased levels of realism for characters in serious or educational games on student experience and learning.

6.5 THE UNCANNY IN HUMANS

Bogart and Matsumoto (2010b) stated that future work should focus on viewer perception of emotion in those with conditions that limit facial movement, such as Moebius syndrome, and how a perceived lack of facial mimicry may disrupt smooth social interaction. "Future studies should examine how others interpret the emotions of people with Moebius syndrome, and how the lack of facial mimicry in people with the condition affects social interaction" (p. 249). Building on this, I put forward that the issues Bogart addressed in her explorations of those with Moebius syndrome extend to viewer perception of the uncanny, not only in human-like virtual characters, but in humans too. With regards to human to human interaction, it has been highlighted that alternative, neurological processes that serve as a compensatory mechanism for those with an inability to mimic others' facial expressions may take a lifetime to develop and even then, the process of emotion recognition may not be as 'seamless' and congruent as in those without such conditions (see Bogart and Matsumoto, 2010b; Goldman and Sripada, 2005). Therefore, this may contribute to a delayed response that interrupts a more efficient, natural flow of reciprocal interaction, to the extent that one perceives a difference or strangeness about that individual. However, what happens when people experience a novel or temporary reduction in facial movement that prevents facial mimicry in adulthood, such as people who have had facial cosmetic procedures such as Botox to reduce movement (and wrinkles) in the face, including the region around the eye, the brows and forehead? Do the same compensatory processes kick in to compensate for a lack of facial mimicry as those who have had congenital facial paralysis conditions do, or would they not have had the time and experience required to develop such skills and compensatory techniques? Empirical research and reports in new media suggest the latter, and cosmetic procedures to limit facial expression may also inhibit social interaction.

As a result of this impaired facial movement, associations of the uncanny have been made with people in society and those in the public eye who have had Botox and facial cosmetic work (see, e.g., Etcoff, 2000; Lam, 2013). Those contributing to an online *Escapist* discussion forum about the Uncanny Valley effect stated that they associated the strange appearance of people who have had Botox with the Uncanny Valley, to the extent that they were repulsed: "After seeing a lot of people use Botox … I am finding myself experience an uncanny valley effect toward the people who use it … The more work people have done pushes them further down the valley to a point of near revulsion." Another likened the appearance of a celebrity to that of a mutant creature: "Joan Rivers, for example, looks like an alien came to Earth, took the skin from someone's face and stretched it over their own head as a disguise" (*Escapist,* 2011, p. 1). Even plastic surgeons are becoming acutely aware of the growing association between the Uncanny Valley phenomenon and those who have had facial cosmetic procedures to enhance or mute parts of the face. On his website, plastic surgeon Dr. Samuel Lam disclosed how patients considering cosmetic procedures are worried that they may simulate the "weird" and "frightening" appearance of others who have had facial work: "One of the biggest concerns that I have from prospective patients is that they not look weird, the fear stemming from seeing unnatural celebrities that are frightening to them in magazines and on television" (2013, p. 1).

However, the question still remains as to *why* we perceive a difference or strangeness as uncanny in those who have had Botox? I propose that, as with human-like virtual characters with aberrant facial expression, it is a perception of a lack of empathy in that Botox individual toward us due to a lack of emotional contagion, to the extent that they may pose a possible threat toward us, evoking the uncanny. Empirical research into emotion recognition in Botox patients conducted by Dr. David Havas et al. (2010), a psychologist at the University of Wisconsin–Madison, offers support for my theory. Havas and his team designed an experiment to assess the role of participants' facial movement and nervous system in their comprehension of emotive sentences. A total of 41 people who wished to have Botox treatments for the first time to reduce their frown lines were recruited to take part in before-and-after experiments (Havas et al., 2010). On the first occasion, before Botox treatment, participants read sentences that described angry, happy and sad scenarios and were then required to press a keyboard button once they had understood each sentence. Example sentences included, "Reeling from the fight with that stubborn bigot, you slam the car door" (anger; p. 896) or, "You hold back your tears as you enter the funeral home" (sadness; p. 896). Then, two weeks after Botox treatment to disable muscles involved with the frown action, including the corrugator supercilii muscle that moves the eyebrows downward and creates deep vertical wrinkles between the brows, participants read the angry, happy and sad questions again. This time participants' responses were found to be significantly slower in reading and understanding the angry and sad sentences (Havas et al., 2010). Botox had impaired not only emotional expression but also emotion recognition and the ability to process emotional language. Without movement of the muscles involved in the frown action, the involuntary, automatic process of simulating emotions such as anger and sadness using facial muscles and nerves was prevented. I suggest that those who have had Botox may not have acquired an efficient and effective kinesthetic (neurological) feedback compensatory mechanism to understand emotion without facial mimicry. As such, they may experience a reduction in their ability to empathize with negative emotional situations, and this may have a detrimental effect on social interaction and how empathetic others perceive them to be. Havas remarked that Botox dampens our ability for emotional contagion: "Botox induces a kind of mild, temporary cognitive blindness to information in the world, social information about the emotions of other people" (quoted in Barron, 2010). This correlates with my explanation of the uncanny in human-like virtual characters with aberrant facial expression in that we fail to see an appropriate response in them to us (and others), which makes us feel uncomfortable. They are perceived as unresponsive and cold, as if blind to another's feelings. Therefore, uncanniness occurs in those who have had Botox due to a perception of a lack of empathy in that person toward us. Emotional contagion and the process of sharing another's thoughts and feelings are qualities that make us human, and if Botox reduces our ability to do this then we may be regarded as *less human-like and stranger*. Paradoxically, people may have Botox with the expectation that they will be perceived as more attractive and likeable without evidence of wrinkles and expression on their face. As I have found with uncanny, human-like virtual characters, however, we may be confused as to their emotional state and alerted to the fact that they cannot recognize our emotion and understand us properly, which means we cannot engage or connect with them. In this way the uncanny may occur on a human-to-human level, based on aberrant facial expression in an individual. Building on the issues raised thus far, in the next chapter I address how a lack of facial mimicry opportunity may prevent us from forging an attachment with a character and may threaten concept of the self and ego.

REFERENCES

Allen, J. F., Byron, D. K., Dzikovska, M., Ferguson, G., Galescu, L. and Stent, A. (2001) "Towards conversational human–computer interaction," *AI Magazine*, vol. 22, no. 4, pp. 27–37.

Barron, J. (2010) "Botox blunts the blues (a little)," *Baseline of Health*. Retrieved February 21, 2014, from http://jonbarron.org/article/botox-blunts-blues-little.

Bell, L. and Gustafson, J. (2003) "Child and adult speaker adaptation during error resolution in a publicly available spoken dialogue system," in *Proceedings of Eurospeech 2003*, Geneva, Switzerland, pp. 613–616.

Blake, R. and Shiffrar, M. (2007) "Perception of human motion," *Annual Review of Psychology*, vol. 58, pp. 47–73.

Blakeslee, S. (2006) "Cells that read minds" *New York Times*. Retrieved February 21, 2014, from http://www.nytimes.com/2006/01/10/science/10mirr.html.

Bogart, K. R. and Matsumoto, D. (2010a) "Living with Moebius syndrome: Adjustment, social competence, and satisfaction with life," *Cleft Palate–Craniofacial Journal*, vol. 47, pp. 134–142.

Bogart, K. R. and Matsumoto, D. (2010b) "Facial mimicry is not necessary to recognize emotion: Facial expression recognition by people with Moebius syndrome," *Social Neuroscience*, vol. 5, no. 2, pp. 241–251.

Busine, A., Abrilian, S. and Martin, J. C. (2004) "Evaluation of multimodal behaviour of embodied agents," in Pelachaud, C. and Ruttkay, Z. (eds.), *From Brows to Trust: Evaluating Embodied Conversational Agents*, Dordrecht, Netherlands: Kluwer Academic Publisher, pp. 217–238.

Carey, B. (2010) "Facial mimicry: The way to express emotions," *New York Times*. Retrieved February 21, 2014, from http://www.deccanherald.com/content/63005/facial-mimicry-way-express-emotions.html.

Corradini, A., Mehta, M. and Robering, K. (2009) "Conversational characters that support interactive play and learning for children," in Ahmed, S. and Karsiti, M. (eds.), *Multiagent Systems*, Croatia: I-Tech Education and Publishing, pp. 349–374.

Darves, C. and Oviatt, S. L. (2002) "Adaptation of users' spoken dialogue patterns in a conversational interface," in *Proceedings of the 7th International Conference on Spoken Language Processing (ICSLP'2002)*, Denver, CO, pp. 561–564.

Darves, C. and Oviatt, S. L. (2004) "Talking to digital fish: Designing effective conversational interfaces for educational software," in Pelachaud, C. and Ruttkay, Z. (eds.), *From Brows to Trust: Evaluating Embodied Conversational Agents*, Dordrecht, Netherlands: Kluwer Academic Publisher, pp. 271–292.

De Weid, M., Van Boxtel, A., Zaalburg, R., Goudena, P. P. and Matthys, W. (2006) "Facial EMG responses to dynamic emotional facial expressions in boys with disruptive behavior disorders," *Journal of Psychiatric Research*, vol. 40, pp. 112–121.

Di Pellegrino, G., Fadiga, L., Fogassi, L., Gallese, V. and Rizzolatti, G. (1992) "Understanding motor events: A neurophysiological study," *Experimental Brain Research*, vol. 91, pp. 176–180.

Escapist (2011) "Poll: Does Botox & plastic surgery produce an 'uncanny valley' effect?" Retrieved February 21, 2014, from http://www.escapistmagazine.com/forums/read/18.287833-Poll-Does-botox-plastic-surgery-produce-an-uncanny-valley-effect.

Etcoff, N. (2000) *Survival of the Prettiest: The Science of Beauty*. New York: Anchor Books.

Gallese, V., Fadiga, L., Fogassi, L. and Rizzolatti, G. (1996) "Action recognition in the premotor cortex," *Brain*, vol. 119, no. 2, pp. 593–609.

Giles, H., Coupland, J. and Coupland, N. (1991) "Accommodation theory: Communication, context, and consequence," in Giles, H., Coupland, J. and Coupland, N. (eds.), *Contexts of Accommodation: Developments in Applied Sociolinguistics*, Cambridge: Cambridge University Press, pp. 1–68.

Goldman, A. I. and Sripada, C. S. (2005) "Simulationist models of face-based emotion recognition," *Cognition*, vol. 94, no. 3, pp. 193–213.

Hatfield, E., Cacioppo, J. T. and Rapson, R. L. (1993) *Emotional Contagion*. Cambridge: Cambridge University Press.

Havas, D., Glenburg, A., Gutowski, K., Lucarelli, M. and Davidson, R. (2010) "Cosmetic use of Botulinum Toxin-A affects processing of emotional language," *Psychological Science*, vol. 21, no. 7, pp. 895–900.

Hess, U. and Blairy, S. (2001) "Facial mimicry and emotional contagion to dynamic emotional facial expressions and their influence on decoding accuracy," *International Journal of Psychophysiology*, vol. 40, 129–141.

Hsee, C. K., Hatfield, E. and Chemtob, C. (1992) "Assessments of the emotional states of others: Conscious judgments versus emotional contagion," *Journal of Social and Clinical Psychology*, vol. 11, pp. 119–128.

Iacoboni, M. (2008) *Mirroring People: The New Science of How We Connect with Others*. New York: Farrar, Straus, and Giroux.

Iacoboni, M. (2009) "Imitation, empathy, and mirror neurons," *Annual Review of Psychology*, vol. 60, pp. 653–670.

Iacoboni, M. and Dapretto, M. (2006) "The mirror neuron system and the consequences of its dysfunction," *Nature Reviews Neuroscience*, vol. 7, pp. 942–951.

Iacoboni, M., Molnar-Szakacs, I., Gallese, V., Buccino, G., Mazziotta, J. C. and Rizzolatti, G. (2005) "Grasping the intentions of others with one's own mirror neuron system," *PLoS Biology*, vol. 3, pp. 529–535.

Jabbi, M., Swart, M. and Keysers, C. (2007) "Empathy for positive and negative emotions in the gustatory cortex," *NeuroImage*, vol. 34, no. 4, pp. 1744–1753.

Keysers, C. (2009) "Mirror neurons," *Current Biology*, vol. 19, no. 21, R971–R973.

Keysers, C. (2011) *The Empathic Brain* [e-book], Social Brain Press or CreateSpace Independent Publishing Platform.

Laird, J. D., Alibozak, T., Davainis, D., Deignan, K., Fontanella, K., Hong, J., et al. (1994) "Individual differences in the effects of spontaneous mimicry on emotional contagion," *Motivation and Emotion*, vol. 18, pp. 231–247.

Lam, S. (2013) "Survival of the prettiest part 3: The uncanny valley," *Lam Facial Plastics: Dr Sam Lam*. Retrieved March 25, 2014, from http://www.lamfacialplastics.com/facial-rejuvenation/survival-of-the-prettiest-part-3-the-uncanny-valley/.

Lipps, T. (1905) "The knowledge of foreign selves" (*Das wissen von fremden ichen*), in Lipps, T. (ed.), *Psychologische Untersuchungen*, vol. 1, no. 4. Leipzig: Engelmann, pp. 697–722.

Lundqvist, L. O. (1995) "Facial EMG reactions to facial expressions: A case of facial emotional contagion?" *Scandinavian Journal of Psychology*, vol. 36, no. 2, pp. 130–141.

McIntosh, D. N., Reichmann-Decker, A., Winkielman, P. and Wilbarger, J. L. (2006) "When the social mirror breaks: Deficits in automatic, but not voluntary, mimicry of emotional facial expressions in autism," *Developmental Science*, vol. 9, no. 3, pp. 295–302.

Oviatt, S. L., Darves, C. and Coulston, R. (2004) "Toward adaptive conversational interfaces: Modeling speech convergence with animated personas," *ACM Transactions on Computer–Human Interaction (TOCHI)*, vol. 11, no. 3, pp. 300–328.

Rizzolatti, G. and Craighero, L. (2005) "Mirror neuron: A neurological approach to empathy," in Changeux, J. P., Damasio, A. R., Singer, W. and Christen, Y. (eds.), *Neurobiology of Human Values*, Heidelberg, Berlin: Springer-Verlag, pp. 107–123.

Rizzolatti, G., Fadiga, L., Gallese, V. and Fogassi, L. (1996) "Premotor cortex and the recognition of motor actions," *Cognitive Brain Research*, vol. 3, pp. 131–141.

Tinwell, A. (2014) "Applying psychological plausibility to the Uncanny Valley phenomenon," in Grimshaw, M. (ed.), *Oxford Handbook of Virtuality*, Oxford: Oxford University Press, pp. 173–186.

Tinwell, A., Abdel Nabi, D. and Charlton, J. (2013) "Perception of psychopathy and the Uncanny Valley in virtual characters," *Computers in Human Behavior*, vol. 29, no. 4, pp. 1617–1625.

Tinwell, A., Grimshaw, M., Williams, A. and Abdel Nabi, D. (2011) "Facial expression of emotion and perception of the Uncanny Valley in virtual characters," *Computers in Human Behavior*, vol. 27, no. 2, pp. 741–749.

Wicker, B., Keysers, C., Plailly, J., Royet, J. P., Gallese, V. and Rizzolatti, G. (2003) "Both of us disgusted in My insula: The common neural basis of seeing and feeling disgust," *Neuron*, vol. 40, pp. 655–664.

Wild, B., Erb, M., Eyb, M., Bartels, M. and Grodd, W. (2003) "Why are smiles contagious? An fMRI study of the interaction between perception of facial affect and facial movements," *Psychiatry Research: Neuroimaging*, vol. 123, pp. 17–36.

Winerman, L. (2005) "The mind's mirror," *Monitor on Psychology*, vol. 36, no. 9, pp. 49–50.

CHAPTER 7

Attachment Theory and Threat to Self-Concept (Ego)

I**N THE LAST CHAPTER**, I discussed how a lack of facial mimicry in a human-like character toward the user and other characters featured in the game or animation may evoke the uncanny due to a perception that the character is unable to understand and empathize with others. However, what happens when we ourselves are denied the opportunity to mimic the facial expression of an "emotionless" human-like, virtual character due to inadequate facial expression? Furthermore, what are the psychological consequences of the realization that a character is not responding to us? Do we just dismiss it, or does it strike a deeper chord with us, in that a character that looks human and who we may expect to behave like a human would is evidently ignoring us? In this chapter I reflect on how a lack of facial mimicry opportunity from a human-like character toward us, and vice versa, prevents us from obtaining self-verifying feedback that challenges our ego and perception of self. From birth, we are primed to forge attachments with others that not only help us develop social skills but are important for our survival and help consolidate a sense of self and who we are (Bowlby, 1951, 1969; Tronick et al., 1975). Based on this, I build the argument for an attachment framework (Tinwell, 2014, p. 180) as a way to understand uncanniness in human-like characters. I consider the role of mirror neuron activity (MNA) and facial mimicry in attachment theory and how a lack of these processes may prevent a viewer from forging an effective attachment with an uncanny, human-like virtual character. Importantly, I consider how aberrant facial expression in a human-like virtual character may cause doubt as to our own social skills and even abnegation of self. This leads to a synopsis of if we are all able to detect uncanniness in human-like characters, or if there may be individuals who cannot detect uncanniness or are less sensitive to this phenomenon based on their mental and physical conditions.

7.1 REFLECTION OF THE SELF

As well as being able to understand another's actions and how they are feeling, MNA and mimicry allow us to identify similarities between ourselves and others. In other words, we use mimicry to emotionally and cognitively empathize with others and to establish how we are akin to them. The neuroscientist Vittorio Gallese, who (as discussed in Chapter 6 of this book) was part of the research team who originally identified mirror neuron cells in macaque monkeys (Di Pellegrino et al., 1992; Gallese et al., 1996; Winerman, 2005), suggested that the mimicry process is not only an important survival mechanism to aid social interaction with others but also can be used to help us grow and prosper in a complex society (Winerman, 2005). Rather than being different, we can approach and communicate with others to find similarities between us, as this helps one gain a better understanding and validity of oneself (Tinwell, 2014; Winerman, 2005). As humans we seek to find similarities between ourselves and others as a means of ensuring smooth interaction with them and also as a means of self-verification for purposes of existential security and psychological coherence (Swann, 1983). To this end, in social situations we expect that we will be able to mimic the facial actions of another, not only to gain a better understanding of their emotional state but also so that we may establish a better understanding of our own identity (Winerman, 2005). As Gallese suggested, in face-to-face situations, the instinctive, automatic process of facial mimicry acts like a social mirror, in that we not only see the person with whom we are communicating but also aspects of ourselves reflected in them: "It seems we're wired to see other people as similar to us, rather than different. At the root, as humans we identify the person we're facing as someone like ourselves" (quoted in Winerman, 2005, p. 48).

We mimic others' facial expression both so that we may empathize with them (and allow them to empathize with us) and as a way to corroborate and evolve concept of oneself. If we see that someone is angry because despite running for a bus they just missed it as it pulled away from the bus stop, we not only experience and understand that person's anger from their frown (and other body gestures) but also can relate to their situation as we may have experienced the same (or similar) situation ourselves. We can understand why they are angry because how they responded resembled how we may have responded. This provides a sense of security in that how we think and feel resembles the thoughts and feelings of others. In this way, our thoughts and feelings are alike with others, and we are more likely to get along with them. This may work to our own advantage in terms of harmonious social interaction to aid survival and so that we are popular and may thrive (Swann, 1983). Professor of Social and Personality Psychology, William B. Swann, Jr., has studied why people formulate and maintain their self-conceptions and create environments that reaffirm the opinion that they hold about themselves (Swann, 1983). Swann explained that, "... *self-verification* processes enable people to create—both in their actual social environments and in their own minds—a social reality that verifies and confirms their self conceptions" (p. 33). Based on this need for self-verification, in society we are more likely to seek out people who respond to us in a way that fits our self-concept and feeds our ego (Swann, 1983; Swann and Pelham, 2002). Hence, we may reject others if we fail to gain self-verifying reactions (e.g., facial feedback) that confirm our own self-views (ibid.). In this way, we use facial mimicry to ensure the stability of our own self-conceptions, for example, our beliefs, preferences and how we perceive that we may react to a given situation. In other words, we want to see ourselves mirrored in others and that people see us as we see ourselves.

So what happens when we interact with a relational character with a highly human-like appearance that not only may lack an ability to mimic our facial expression but also may prevent us from being able to mimic and understand their emotional state (i.e., to engage with that character)? We may expect to see similarities between our behavior and theirs, so how do we respond to disparities between our verbal and nonverbal behavior and that of the character? I put forward that, when confronted with a virtual character with a realistic, human-like appearance but with inadequate facial expression, as well as experiencing confusion as to that character's emotive state (so that the future actions of that character are less predictable) one may also perceive that oneself is not being recognized as human or understood by that character (Tinwell, 2014). The social interaction with this character does not run smoothly and is sent off track from what we would expect. Therefore, not only is self-verification not achievable, but also one's ego may be bruised. A perceived lack of mimicry response in a human-like character to one's own emotions or an inability for us to mimic the character may lead to essentially an existential angst, with no reflection of oneself shown in the character's face (Tinwell, 2014). However, where does our ego come from and how does it drive this anxiety and dread when interacting with an uncanny, human-like character? In the following sections I consider what our ego and identity stand for. Furthermore, I consider how a perceived lack of facial mimicry and affective reciprocity from a human-like character may instigate a detachment reaction in the viewer that deviates from infancy and may cause doubt as to one's own existence.

7.2 SELF, IDENTITY AND ATTACHMENT THEORY

As part of this investigation into why we experience the uncanny in human-like virtual characters, I am compelled to consider the ontological (and psychological) status of the self. What constitutes a sense of self, and where is this derived from? Questions such as these have tested psychologists and philosophers for centuries and, as previous researchers have acknowledged, "From the beginning, psychology's relationship with the 'self' has been a tempestuous one" (Swann and Bosson, 2010, p. 589). The notion of the self was first introduced in the late twentieth century by the psychologist and philosopher William James (1842–1910) and may be explained as the cognitive, affective and social behavior that forms one's personality and identity (Swann and Bosson, 2010; James, 1890a, 1890b). In his essay "The Self and Its Selves" (James, 1890a), James reflected on how both conscious and subconscious processes in our thoughts, feelings and behavior help us to establish our sense of self. For example, we may rely on our semantic memory system to reinforce our qualities and traits with thoughts such as, "I always stick at a job to get it finished and don't give up," or to reinforce perceptions of our social roles and traits such as, "I am a friendly person" (Swann and Bosson, 2010). Importantly, James suggested that the construct of the self provides us with a sense of stability: "James saw the self as a source of continuity that gave the individual a sense of 'connectedness' and 'unbrokenness'" (Swann and Bosson, 2010, p. 589). The ego has been described as one's inner voice (Frederick, 2013, p. 39) that may act as our guide and caution us against unwise decisions and potentially harmful situations (Dyson, 1998; Frederick, 2013). Two binarisms have been proposed, namely, the *I* and the *me* that constitute oneself (James, 1890a; Swann and Bosson, 2010). The *I* is based on one's subjective interpretation and experience of the self, and the *me* is a more objective, physical object (James, 1890a). A sense of stability and continuity between our previous, present and impending selves may be tied together in our ego (James, 1890a, 1890b). James theorized that the ego represents our implicit thoughts about ourselves and our identity that can help us gain a more positive outlook about ourselves (ibid.). We build upon our past experiences, present state and what we predict we will be like in the future to establish a sense of self and ego. In this way, the ego helps to build our self-esteem, pride and self-love, even to the extent that one may be perceived as self-absorbed and selfish (ibid.). Our ego lets us project our best possible self in our own minds and how others see and respond to us (Swann and Bosson, 2010). As a way to feed and enhance our ego, we like to seek out objects and people that remind us of ourselves. Choices that we make about our line of work, partner, friends and material objects such as our house and car may be guided by our identity and ego (Pelham, Carvallo and Jones, 2005; Swann and Bosson, 2010). These objects and the people that we choose to socialize with represent our self-worth. Given that most people have a more positive interpretation of the self, we will automatically look for things that resemble the self, and we may regard people and objects that do not remind us of ourselves less favorably (ibid.).

The facets of the self and their underlying phenomenological and neural substrates attract continued interest by theorists and scientists to help define when and how we build our own identity (ego) and model of the self (Swann and Bosson, 2010). To help understand when we may develop a sense of self, psychologists have looked to the work of ethologists and observations made with animals. The notion of imprinting in animals suggested that, at birth, there is a critical period within which animals will form a crucial attachment with the first moving thing that is visible to them. Evidence of this can be found in ducklings that upon hatching will form an irreversible, unbreakable bond with the first animal (or person) that they see (Bowlby, 1969; Lorenz, 1935, 1952). To demonstrate this, in 1935 animal behaviorist Konrad Lorenz divided unhatched goose eggs into two groups: one group that was hatched by their original mother; and another incubated group so that he would be the first thing that the ducklings would see after hatching. Once hatched, the incubated group first saw Lorenz and thus became dependent on him. The ducklings followed Lorenz everywhere and, unable to reverse their fixed, rigid bond with him, became distressed if they could not find him (Bowlby, 1969; Lorenz, 1935, 1952). Whether it is their mother goose or a human, before ducklings can fledge and become independent, they are reliant on their caregiver to provide for them and to help them survive. This innate pattern of behavior has advantages to both the offspring (e.g., a duckling) and parent (e.g., a mother goose). The young will be kept safe and warm, provided with food and taught how to fend for themselves. It is also in the parent's best interest to see their offspring survive and go on to reproduce. This not only rewards the time that a parent has spent on their offspring before and after birth but also ensures the success and evolution of their species (Lorenz, 1935, 1952).

Building on this work with animals and the notion of imprinting, the psychologist and psychoanalyst Edward John Mostyn Bowlby (1907–1990) proposed the notion of attachment theory in humans (Bowlby, 1969). As shown in animals, from birth we instinctively bond with others as a means of survival (Bowlby, 1969; Hazan and Shaver, 1994). Bowlby (1969) suggested that we have an innate instinct to display behaviors that help ensure and maintain social contact with others. Infants seek to forge attachments and use behaviors such as crying, smiling, gurgling and crawling to gain attention and achieve a closer relationship with their caregiver (Bowlby, 1969). Crying, smiling and other gestures and movement are used as signals to the caregiver, who in turn should respond with animacy and expression to signal to the child that they have been noticed. This responsive reaction from the caregiver helps to create a reciprocal, appropriate and balanced means of interaction with the infant. Evidence has shown that the first two years of an infant's life are crucial for developing sound and successful attachments (Bowlby, 1951, 1969; Swann and Bosson, 2010; Tronick et al., 1975). The attachments that we forge in these early stages influence development of our self-awareness and are fundamental in how we may ultimately forge attachments in later life (Bowlby, 1951, 1969; Bowlby and Robertson, 1952; Swann and Bosson, 2010). It is in this early stage of infancy that we start to build basic models about our self-worth and how loveable we are (Bowlby, 1951, 1969; Swann and Bosson, 2010). These working models that we hold about ourselves are affected by the treatment that we receive from our caregivers and our responsiveness to that treatment (Bowlby, 1951, 1969; Bowlby and Robertson, 1952; Swann and Bosson, 2010). Optimal development occurs when there is an affect harmony between the infant and caregiver (Bowlby, 1969; Prior and Glaser, 2006; Tronick et al., 1975). To convince an infant that they are worthy of love and affection and are capable of forging effective and successful relationships requires responsive and consistent caregiving to that infant (Bowlby, 1951, 1969; Swann and Bosson, 2010). Positive and responsive attachments forged in infancy help build the foundations of a high self-esteem, a positive self-concept and self-assuredness (Bretherton, 1988; Swann and Bosson, 2010). In turn, this kinship inspires self-confidence and social skills in that child, both of which help to improve their future potential to forge successful relationships and to aid survival.

However, negative or unresponsive interactions that occur in early life can have a long-term detrimental effect on our ability and potential to develop effective interpersonal communication skills (Bowlby, 1951, 1969) and also prevent us from building a more positive self-concept (Bowlby 1951, 1969; Swann and Bosson, 2010; Tronick et al., 1975). If there is a lack of caregiving and a child is neglected or if there is inconsistent, unresponsive or abusive interaction with a child, this may result in a more negative view of how they see themselves and others (Bowlby, 1951, 1969; Swann and Bosson, 2010; Tronick et al., 1975). Researchers have observed that the attachment patterns that we acquire in early life stay with us to adulthood (Bowlby, 1951, 1969; Fonagy, Steele and Steele, 1991; Hesse, 1996). Therefore, such negative interaction may instill doctrines that one is not valuable as a person and that others are untrustworthy and cannot be relied upon. This results in low self-esteem and a low regard for oneself (Bowlby, 1951, 1969; Swann and Bosson, 2010; Tronick et al., 1975). Interestingly, Bowlby (1951, 1969) proposed that many dysfunctional, maladaptive mental and behavioral disorders could be traced back to a lack of effective, responsive attachments in infancy. An individual may suffer from depression and demonstrate a reduced intelligence, increased aggression and antisocial behavior due to social deprivation in infancy. These long term psychological, social and emotional difficulties may even be to the extent of affectionless psychopathy (Bowlby, 1951, 1969). Individuals diagnosed with affectionless psychopathy show a lack of concern or care for others and an inability to forge relationships, which Bowlby (1951, 1969) attributed to a lack of care and successful attachments in infancy. However, as I discuss in the next section, we are equipped with behaviors to try to prevent us from being exposed to unresponsive interactions.

7.3 PROTEST, DESPAIR AND DETACHMENT BEHAVIOR

If an infant senses a lack of attention and empathy from a caregiver, then as a way to try to prevent this they are inclined to show a protest–despair–detachment pattern of behavior (Ainsworth et al., 1978; Bowlby and Robertson, 1952; Holmes, 2001). An infant will initially make a strong protest against the caregiver to signal their dissatisfaction at a lack of attention from them. This protest may include behaviors such as loud shrieking and shaking of their head as a way to try to evoke a response from their caregiver. Following this, the infant may portray sadness toward the unresponsive caregiver with crying and whimpering. If these tactics fail, then the infant may ultimately reject and detach from their caregiver as they have learned that attempts to make responsive contact are ignored (ibid.). If this occurs on a continued basis, then this may be potentially damaging for that infant as they are at an overt risk of developing dysfunctional attachment behavior. Such dysfunctional characteristics may work against one's future ability to forge successful relationships with others that are important for one's survival (Bowlby, 1951; Fonagy et al., 1991; Hesse, 1996; Tronick et al., 1975).

In 1975, the psychologist Edward Tronick (an infant–parent mental health expert) and his colleagues designed an experiment to demonstrate how responsive infants are to social influence from others and the consequences of a lack of responsiveness from their caregiver (Tronick et al., 1975). As head of the Child Development Unit at the University of Massachusetts, Tronick is renowned for his research into social, emotional development in infancy and early childhood (Infant-Parent Mental Health Expert Ed Tronick, 2013, p. 1). Tronick wanted to show that infants could engage in social interaction and relied on facial expression to do so. The experiment was called "The Still Face Experiment" (Tronick et al., 1975; UMassBoston, 2009) and involved a mother and her 12-month-old baby as participants. In the first stage of the experiment, the mother and her baby sit facing one another, and the mother begins to play with the baby. The mother greets the baby, and the baby returns his mother's greeting, as they acknowledge each other's presence in a friendly way (UMassBoston, 2009). The baby uses gestures such as pointing at objects in the room. Accordingly, the mother responds by smiling and showing interest, enthusiasm and attention to her baby and the object her baby is pointing at. The mother mimics her baby's facial expression so that when the baby smiles, the mother smiles too, and vice versa. In this way they establish an affect harmony as their emotions and intentions mirror one another. The baby is used to playing with its mother and receiving her full attention and a continued level of interest and responsiveness from her. Tronick says that "they are working to coordinate their emotions and their intentions, what they want to do, and that's really what the baby is used to" (UMassBoston, 2009). Then, in the second stage of the experiment the mother is asked not to respond to her baby. The mother turns her head away from her child so that the infant is faced with the back of his mother's head. Then, when the mother turns around to face her baby again, she has a blank, emotionless expression (UMassBoston, 2009). The baby immediately notices his mother's blank expression, so he adapts its behavior to try to get the mother's attention again. With wide, open eyes and raised brows the baby smiles at the mother to signal interest and to try to gain closeness with her. As the mother's face remains unmoved and still with no response, the baby, who is used to his mother responding to him, points at objects in the room. At the same time the baby also raises his eyebrows to signify interest, as if to say, "Look!" (UMassBoston, 2009). With yet again no response from his mother's blank face, the baby starts to show a protest–despair–detachment response. The baby's eyebrows lower to show a frown expression to communicate that he is growing increasingly angry and frustrated at his mother. The baby leans closer to his mother's blank face and raises both hands as if to ask, "What is going on?" (UMassBoston, 2009). Given that the baby is used to his mother showing mirrored facial expressions and a reciprocal, harmonious response, the baby is confused and panicked by his mother's blank, emotionless face. In exasperation at still no response from his mother's blank face, he makes a high-pitched shrieking sound to try to bring his mother back (UMassBoston, 2009). The baby then becomes upset and starts to cry and wail. The sides of the baby's mouth droop, and his eyebrows lower to show a sad, anguished expression. The baby eventually reaches the point where he can no longer look at his mother's blank face as he has found the situation too stressful, so he turns his head away from his mother. As the baby cries, he uses his hands to cover his face to show that he is becoming detached from and wants to reject this unresponsive person (UMassBoston, 2009). After two minutes of the baby being presented with a blank face from his mother, the mother then returns to her usual responsive self and smiles and comforts her baby, who is in a distressed state. The baby returns to face the mother again and quickly responds to her happy facial expression and emotion by returning the mother's smile. The mother and baby join hands, and their heads get closer as they return to a more positive, emotional interaction (UMassBoston, 2009).

In Tronick et al.'s (1975) experiment, the mother deprived her baby of any facial expression and response for just two minutes. However, this was still long enough to elicit a negative, protest–despair–detachment response in the infant. As a way to explain what had happened during the experiment, Tronick stated that this study had demonstrated the "good, bad, and ugly states" (UMassBoston, 2009) that can occur in child social interaction. Tronick described the good as normal, continuous responsive interaction with an infant: "the normal stuff that goes on, that we all do with our kids" (UMassBoston, 2009). The bad is when a bad situation may occur, with a lack of responsiveness from a caregiver, but on a momentary or temporary basis (Tronick et al., 1975; UMassBoston, 2009). Even though the bad situation has occurred, because it was just on a short-term basis the infant can recover from it and return to a good situation again. As Tronick stated, "After all, when we stopped the still face, the mother and the baby start to play again" (UMassBoston, 2009). However, an ugly situation may occur if an infant is continuously left for long periods of time, without any positive, affective social interaction (Tronick et al., 1975; UMassBoston, 2009). The ugly is when an infant is not given a chance to return to a good situation and is left deprived of positive interaction (ibid.). With no opportunity to make amends, they are suspended in that ugly situation (ibid.). Overall, Tronick's experiment showed that infants are highly responsive to facial expression of emotion and try to encourage good situations and reciprocal, meaningful interaction with others (Tronick et al., 1975). Importantly, he stated that if exposed to an ugly situation, this may cause an infant long-term harm to their self-confidence and social, emotional development.

> An infant's exposure to "good, bad and ugly" interactions with the mother, as repeatedly communicated by her facial expressions or lack of expression (i.e., a still face) has long-term consequences for the infant's confidence and curiosity, or social emotional development, with which to experience and engage the world. (Tronick, 2013, p. 1)

Tronick's work helped support Bowlby's (1951, 1969) previous theory, in that adults who have experienced ugly situations in infancy may lack self-belief and be more predisposed to develop antisocial tendencies, (possibly to the extent of affective psychopathy) in later life. But what does the notion of good, bad and ugly situations (Tronick et al., 1975; UMassBoston, 2009) mean in the context of the uncanny in human-like virtual characters? Tronick's experiment (and the work of other researchers such as Bowlby, 1969 and Swann, 1983) infers that we rely on an acknowledgment and recognition of self from others via their facial expression to affirm and corroborate our own identity (ego) and model of self. From this perspective, we use other people's response to us and their facial expression as a sign to confirm that we exist as a person (Bowlby, 1969; Swann, 1983). Tronick and his colleagues (1975) predicted that ugly, undisturbed periods of a lack of social interaction may result in irrevocable antisocial behavior in an individual. As shown in Tronick's experiment (see also UMassBoston, 2009), babies can transfer from bad to good situations with the opportunity for reparation as long as they receive normal (i.e., good), attentive and responsive periods of interaction with their caregiver. Without this vital interruption of good situations, there remains the permanent risk of developing bad socialization behavior. Building on this and the negative symptoms that the baby in Tronick's "Still Face Experiment" (Tronick et al., 1975; UMassBoston, 2009) showed in being exposed to a bad situation when he saw his mother's blank face, I suggest that we experience the uncanny as a way to prevent us from being at risk of a bad or ugly situation that may have a detrimental effect on our own social skills and well-being (Tinwell, 2014). In this sense, uncanniness alerts us to the danger of a potentially ugly and self-destructive situation that, if exposed to an emotionally inert human-like character to a great extent, could negate our self-concept and social, emotional health (Tinwell, 2014). I propose that, with no evident signs of facial expression of emotion from its mother, the baby could not detect any mimicry of its own facial expressions in its mother (Tinwell, 2014). Furthermore, the baby was prevented from being able to mimic its mother's facial expressions. This situation may not have only caused distress for the baby in that they were not being recognized or acknowledged by their mother but also impeded the baby's developing awareness of self, hence eliciting the uncanny (Tinwell, 2014). Given that the learned attachment (and detachment) patterns of behavior in infancy can remain intact into adulthood (Fonagy et al., 1991; Hesse, 1996), I propose that this condition is symptomatic of the cause of humans experiencing the uncanny when interacting with emotionally limited, realistic, human-like virtual characters. Therefore, continued and frequent interaction with unresponsive, human-like virtual characters with inadequate facial expression may be perceived as a threat to one's own well-being and we may disengage with the character to protect and preserve self (Tinwell, 2014).

Traumatic memories such as being exposed to an unresponsive caregiver in infancy can stay with us, hidden in the subconscious, subcortical regions of the brain (Barabasz, 2013; van der Kolk, 1994). As well as short-term memories, the hippocampus is involved in organizing and storing our long-term memories (Moser and Moser, 1999; Squire, 1992; Wood, Dudchenko and Eichenbaum, 1999). Long-term memories that may have caused distress or trauma can stay stored away where they are not readily accessible by the thinking, comprehension and rational parts of the brain in the frontal lobes (Barabasz, 2013; van der Kolk, 1994). However, certain objects or circumstances can cause flashbacks and remind us of these hidden memories that heighten attention and arousal, with particular neurobiological responses (ibid.). As such, we can relive in the present past experiences that we may have thought that we had forgotten (ibid.). Based on this, I propose that when we disengage with and reject an uncanny human-like character, this reaction is prompted by a protest–despair–detachment response that we acquired in infancy (Tinwell, 2014). When adults experience uncanniness, we are actually referring back to the ingrained memories that we hold of the protest–despair–detachment behaviors that we employed in infancy to help us avoid bad or ugly situations. Even if these bad situations that we may have encountered were very brief, we now resort to the same tactics when we are faced with an unresponsive, human-like virtual character as we would have done if we sensed a lack of attention and affect harmony with our childhood caregivers. When we fail to establish an affinity with a human-like character due to their inadequate facial expression that shows no mimicry and reflection of self, we risk being in an affect disharmony or dispute with that character (Tinwell, 2014). Given the distress and anguish we experienced as infants during a protest–despair–detachment response, as adults this inherent memory may make us feel agitated and uncomfortable so that we reject the strange, uncanny character. Rather than the conscious parts of the brain, it may be our subconscious memories and our reaction to them that prompts experience of the uncanny. Therefore, when we attempt to interact with an emotionless, unresponsive human-like character, we are ripe to experience possible flashbacks and retraumatization that triggers a protest–despair–detachment reaction and hence enter optimum conditions for experience of the uncanny. From this perspective, the Uncanny Valley is related to issues of survival, but not necessarily as a reminder of one's own death, as suggested by previous authors (see, e.g., MacDorman and Ishiguro, 2006; Mori, 2012). Rather, experience of the uncanny serves as "as an adaptive alarm bell to remind the person of the importance of being able to form attachments with others; a necessary survival technique to avoid death" (Tinwell, 2014, p. 181). In other words, we rely on experience of the uncanny to alert us to a potentially bad or ugly situation that we should avoid to protect our own healthy social, emotional disposition and perception of self from potential harm (Tinwell, 2014).

7.4 THREAT TO SELF-CONCEPT (EGO)

At this point, it is prudent to assert that if we didn't refer to and enact this protest–despair–detachment reaction to unresponsive human-like virtual characters, then what would the alternative be? Other researchers have proposed that retraumatization of hidden long-term memories and their associated psychophysiological response can be related to an instinctive ego defense mechanism (Barabasz, 2013; Barabasz and Christensen, 2006; Christensen, Barabasz and Barabasz, 2009). We maintain these memories to alert us to possible future circumstances that may be harmful to us and our self-concept so that we may avoid them. Therefore, what impact would frequent and prolonged exposure to human-like characters who may not always show appropriate and reciprocal facial expression to us have on our psyche and how might we rationalize this odd, unexpected behavior? Might this threaten our ego and suggest a divulgence of something wrong with oneself? How would our inner voice (Dyson, 1998; Frederick, 2013) respond to such a strange and disturbing situation? I now discuss how we may interpret abnormal interaction with an unresponsive character as the possibility of aberrant, maladaptive social skills in ourselves. Rather than something being wrong with the character, we may momentarily question, "Is it me?" and perceive that something is wrong with us that may escalate to the morbid realization that one may no longer exist.

Collectively, evidence suggests that verification of our own existence may be affirmed by others showing recognition to us via facial expression that helps us confirm our own identity and perception of self (Bowlby, 1969; Swann, 1983; Tronick et al., 1975). In this respect, we look to forge meaningful attachments with others to reinforce our own identity and feed our ego (Pelham et al., 2005; Prior and Glaser 2006). Therefore, when we are presented with a character with a realistic, human-like appearance, we may perceive the possibility of forging an attachment with that character (Tinwell, 2014). A high level of graphical realism in that character's appearance may raise our expectations that the character will respond to us as a human would. Yet, as has been repeatedly observed and measured in such human-like characters, limitations in the characters' facial expression and behavior may not match their more sophisticated human-like appearance (see, e.g., Corradini, Mehta and Robering, 2009; Tinwell et al., 2011; Tinwell, Abdel Nabi and Charlton, 2013; Vinayagamoorthy, Steed and Slater, 2005). If we smile at a character yet they do not return a friendly expression, what are we left to think? I put forward that, given that we rely upon the facial expression of others to mirror and validate self, doubts may be triggered as to oneself and existence (Tinwell, 2014). Not only may we be reminded of the anguish and more negative behaviors that we may have experienced in infancy when confronted with an unresponsive caregiver, but this uncanny interaction also is a threat to our own ego and self-concept (Tinwell, 2014). If there is a lack of mimicry and affect harmony in a realistic, human-like character, despite our efforts to communicate with that character, then we may question why we are not being realized as a human: "Without this affirming exchange of nonverbal interplay and communication, either verbal or nonverbal, the certainty of one's own existence may be questioned" (p. 181). If we cannot see representation of our physical self in that character's facial expression (or by any other means such as gesture, speech tone and prosody) then this goes against our personal experience and beliefs. This is disconcerting as we may be putting ourselves at risk of developing bad socialization skills ourselves and it threatens our established, more positive self-concept that we hold. Even if this motor mimicry occurs swiftly with only subtle, transient changes in another person's facial expression (Hatfield, Rapson and Le, 2009; Lundqvist, 1995), a total lack of observable change in a character's facial expression to simulate one's own may result in panic or annoyance, preventing us from engaging with that character. This may lead to more alarming ramifications such as a fear of the demise or destruction of the self in that (1) one is apparently not receiving an emotional response from the character and (2) one is unable to mimic and interpret the facial actions of that character, suggesting possible abnormal social skills in oneself (Tinwell, 2014). We may come to the conclusion that the subjective *I* or the physical, objective *me* (James, 1890a) does not exist or is less effective than usual as it failed to prompt a response from another (Tinwell, 2014). There may be the realization of a loss of the ability to mimic another's facial expression and in this way understand their thoughts and feelings, skills that are crucial for effective social interaction. If we cannot mimic and understand the facial expressions and emotions of others, then we may be unable to predict the future actions of others nor establish or maintain effective relationships with them, both of which are important to one's survival (Fonagy et al., 1991; Hesse, 1996). Filled with consternation, this instills doubt of self and the possible abolition or death of self (Tinwell, 2014).

A suggestion or admission of abnormal social skills in one's self is in keeping with Freud's (1919) analogy that the uncanny is elicited by a revelation of the repressed. The earlier more uncomfortable experiences of bad (or ugly) social situations and our response to them that may have been repressed have now come to light. Furthermore, rather than the uncanny being a revelation of something that is wrong with the human-like character, the character's aberrant facial expression has brutally exposed the possibility of negative or maladaptive social skills within one's self (Tinwell, 2014). In the viewer's mind they may perceive that, "I always presumed that I had good social skills and was well received and liked by others, yet this human-like character does not even realize that I am here. Am I to blame and if so, what is wrong with me? Am I really here at all?" Unpleasant thoughts such as these may distort or damage one's perception of self, thus causing alarm and panic in us. A lack of nonverbal communication in a character's upper face suggests a lack of empathy toward us (Tinwell et al., 2013), yet this unexpected, unresponsive interaction may be caused not by the character but by our lack of social skills. Such a situation infers the probability of antisocial, dysfunctional behavior in oneself. However, our ego may not allow this negative doubt of self to subsist and immediately rationalizes this strange interaction as the character's fault and not our own. Our inner voice may step in and conclude, "This character is behaving oddly and I am not comfortable in this situation. The character is strange and creepy, and I am no longer going to tolerate this." Hence, we may automatically reject those human-like virtual characters who offer no evidence of facilitated reciprocal mimicry or self-verifying feedback as a protection mechanism, in self-defense of our ego and well being (Tinwell, 2014). Otherwise, we may be forced to accept that there is actually something wrong with us. Rather than entertain the idea that we may possess abnormal social skills that prevent us from mimicry and effective social interaction with the human-like character, we blame the "strange" uncanny character instead and reject them for their odd behavior (Tinwell, 2014). In summary, uncanniness occurs by a lack of mirror-mimicry of one's self in a human-like virtual character that inflicts fear and dread at the possibility of death of self, so we refer back to a protest–despair–detachment response to interrupt this disruptive and harmful situation. I suggest that this angst and rejection manifests as the Uncanny Valley phenomenon (Tinwell, 2014).

7.5 OBJECTIVE QUANTIFICATION OF UNCANNINESS AND FUTURE WORK

Future experiments may be conducted to test my theory that a lack of facial mimicry response in a human-like character toward us and a lack of opportunity for us to mimic a character's facial expressions acts as an underlying cause of the uncanny. So far, the experiments that I have conducted to assess uncanniness in human-like virtual characters have relied upon subjective participant feedback from questionnaires. However, physiological data would be required alongside the psychological data to help test the theory of a lack of facial mimicry response (and mimicry opportunity) on perception of the uncanny in human-like characters (Tinwell, 2014). This may help improve the validity of results in experiments while, importantly, the relationship between objective and subjective ratings of the uncanny in characters could also be assessed. As well as participants rating on a numeric scale how much they agree or not that a character is strange (and for other variables associated with the uncanny such as whether a character is perceived as cold, callous or unresponsive or not), physiological data should be collected from participants using facial electromyography techniques (EMG). EMG is a technique that measures electrical charge from facial muscle contraction, so mimicry response in participants could be monitored as they interacted with a human, a fully animated human-like virtual character and a human-like character with a lack of upper facial movement. The following outcomes may be predicted: first, the human-like character with a lack of upper facial movement would be rated as most uncanny compared to the fully animated character or human; and, second, participants would also demonstrate less facial musculature activity in response to the partially animated character when compared with the other stimuli. Obtaining these EMG readings as a more objective measure of uncanniness may help explain further why people regard characters with aberrant facial expression as uncanny.

However, empirical evidence investigating the effect of MNA in response to androids offers an alternative explanation to my second hypothesis. Rather than reduced or a lack of facial mimicry response in the viewer toward the partially animated character with no movement in its upper face, there may be an increased level of facial movement in the viewer as they attempt to make sense of this character's unexpected and limited response. In 2012, Ayse Pinar Saygin, associate professor in the Cognitive Neuroscience and Neuropsychology Lab at the University of California, led an experiment to test how our mirror neuron system responded to moving stimuli presented with varying levels of human-likeness. Saygin and her team performed functional magnetic resonance imaging (fMRI) on participants as they watched short video clips of a human, a robot with a mechanical, metallic appearance or an android undertaking everyday tasks such as waving, nodding and drinking a glass of water (Saygin et al., 2012). The android used in Saygin's experiment was named Repliee Q2 and was developed at the Intelligent Robotics Laboratory at Osaka University with professor Hiroshi Ishiguro (2006). As a simulacrum of a woman, Repliee Q2 has the ability to move its head and upper body and has multiple modes of movement (degrees of freedom). With a highly human-like appearance, Repliee Q2 has passed the Turing test, whereby viewers are unable to distinguish Repliee Q2 from a human when viewed at certain distances and for brief periods (Ishiguro, 2005, 2006; Saygin et al., 2012). Named after the artificial intelligence theorist Alan Turing, the Turing test determines if a man-made machine can be distinguished from a human and was initially pioneered to assess natural-language dialogue between human and machine (Turing, 1950). When participants observed Repliee Q2 presenting micromovements (Ishiguro, 2006, p. 327), similar to those slight, involuntarily movements in humans such as a blink of the eyes, for just two seconds duration, a significant majority (77% of participants) mistook the Repliee Q2 for a human. However, as Ishiguro acknowledged, fewer participants may have made this judgment if the android demonstrated a fuller range of motion. However, in its current implementation, the motion capabilities of Repliee Q2 fail to simulate authentic and believable human movement (Pollick, 2010; Saygin et al., 2012). As such, Repliee Q2 has failed the Turing test with longer periods of exposure and when movement was part of the test (ibid.).

Given these limitations in Repliee Q2's movement, Saygin wished to investigate how our mirror neuron system would react to this anomaly of an android with a highly human-like appearance but with limited movement when compared with a mechanical robot and a human (Saygin et al., 2012). So that the mechanical movements would be similar for the mechanical robot and Repliee Q2, the mechanical robot that Saygin and her team used was Repliee Q2 stripped of the human-like features such as synthetic muscle, skin and hair (Saygin et al., 2012). The human featured in the video was the woman that Repliee Q2 replicated (Saygin et al., 2012). Saygin's team analyzed fMRI data from 25 healthy participants aged between 20 and 36 years old and revealed a (seemingly paradoxical) significant, heightened level of MNA when viewing Repliee Q2. This finding was surprising in that, rather than there being reduced MNA in the temporal, parietal and frontal areas of the brain due to a lack of comprehension of what the android was doing and was about to do next, this uncertainty actually increased MNA. As an explanation for this, Saygin and her team proposed that the brain could not process the incongruity between Repliee Q2's robotic motion and its human-like appearance. When we view an object, we expect that there will be accord between that object's appearance and its motion, whether that is a mechanical appearance with jerky motion, or a human-like appearance with biological motion (Rao and Ballard, 1999; Saygin et al., 2012). Accordingly, there was a normal, predicted level of MNA in participants for the human video and mechanical robot. The jerky movements of the robot matched its mechanical appearance and the human's natural movements were expected based on its natural, biological appearance (Saygin et al., 2012). However, once conflict was identified between an object's appearance and motion, then prediction error was increased. Heightened levels of confusion occurred when the viewer was presented with Repliee Q2 with a human-like appearance but with jerky motion, which resulted in a greater prediction error as the brain attempted to understand what was going on. In this way, Repliee Q2's jerky, unnatural motion had caused a "violation of the brain's predictions" (pp. 414–415), and this disharmony may contribute to the uncanny in androids (Saygin et al., 2012). Based on Saygin et al.'s work, rather than reduced (or no) MNA and facial mimicry, a perception of abnormal or a lack of facial expression in a human-like character may evoke increased MNA and facial movement in a viewer attempting to make sense of the incoming socioemotional information. The actions and emotions of the character are not predictable, thus causing a proliferation of neural and muscular activity. Such unpredictability and confusion may be an uncomfortable (even stressful) state for the viewer contributing to the uncanny. Given the outcomes from the experiment by Saygin et al., it may be the case that in my future work, participants may show an increased level of facial musculature activity when presented with human-like virtual characters with aberrant facial expression due to an increased prediction error. As well as EMG, participants' cognitive responses may also be measured using fMRI techniques to establish if MNA (and prediction error) may be decreased or increased when viewing and/or interacting with human-like virtual characters with a lack of facial movement and emotional expressivity. Adopting techniques such as EMG and fMRI in the exploration of the uncanny in human-like characters will help quantify the uncanny as a tangible, objective response in humans, beyond a more speculative, subjective analogy (Tinwell, 2014).

7.6 DO WE ALL EXPERIENCE THE UNCANNY IN HUMAN-LIKE CHARACTERS?

Starting with the assumption that experience of the uncanny is due to a perceived lack of discernible empathy in a human-like character toward others, and that facial mimicry and MNA are important factors in emotional (and cognitive) empathy, I put forward that there may be individuals with physical and/or psychiatric conditions that prevent them from perception and experience of the uncanny. For example, those with conditions that cause a lack of ability for facial mimicry and to empathize with others may be less sensitive to the uncanny than those not diagnosed with such conditions. This may include individuals that cannot move their facial muscles to create a mimicry response, such as those with Moebius syndrome; also, those on the autistic spectrum who do not rely upon a facial mimicry response in social interaction. Furthermore, those diagnosed with antisocial and disruptive behavior disorders and psychopathy that may have an impaired empathic response to others may also be less sensitive to the uncanny. To explore this theory that some might be unable to detect the uncanny in human-like virtual characters, I shall look to previous studies to determine the characteristics of the aforementioned conditions.

Given that those diagnosed with Moebius syndrome have facial paralysis that prevents them from creating and mimicking facial expressions (Bogart and Matsumoto, 2010a, 2010b; Briegel, 2006) it may be that they are less perceptive of a human-like virtual character's lack of emotional expressivity toward others and themselves. This in itself may be enough to reduce (or remove) sensitivity to the uncanny, but there still may be other factors such as a character's jerky motion and speech qualities that alert them to a perceived strangeness in a human-like character. Previous studies have revealed that those with Moebius syndrome can rely on speech, body movements and gesture to help understand another's emotive state (Bogart and Matsumoto, 2010a), so they may detect anomalies in a human-like character's body movements and speech that still render the character as uncanny. Similarly, those diagnosed with autism have been shown not to present a facial mimicry response to others during social interaction

(McIntosh et al., 2006). This may in part contribute to a lack of awareness of another's emotive state (Bogart and Matsumoto, 2010b; McIntosh et al., 2006). In this sense, those on the autistic spectrum may be less sensitive to a lack of emotion and responsiveness in a human-like virtual character as they do not rely on such facial markers to aid social interaction. Without the subconscious, automatic process of mimicry to understand another's facial expression, those with autism can refer to other regions of the brain involved with conscious attention and intentional recognition to help comprehend another's facial expression (ibid.). As I suggested in Chapter 6 of this book, it may be that when those not diagnosed with autism interact with an emotionless human-like character, they too are forced to rely upon conscious and intentional reasoning as a compensatory mechanism to try to understand the character's emotional state. Having to refer to one's intentional thought processes may be regarded as a strange and unfamiliar process as one may be more accustomed to an automatic, thoughtless process of emotion recognition. Hence, the uncanny is exaggerated for a character. However, those with autism may be used to this intentional process of emotion recognition and be less discerning of another's emotional state in the real world. Therefore, these characteristics would make them less sensitive to uncanniness in human-like characters. Measuring sensitivity to the uncanny in human-like characters for Moebius syndrome and autism groups remains an area for future investigation.

Those with antisocial tendencies have a lack of empathic concern for others and clinical studies have revealed that a lack of facial mimicry response may contribute to this lack of empathy. As well as hostile and defiant behavior toward others, children and adolescents with disruptive behavior disorders (DBD) have a weak ability to understand and share the emotional state of others (Cohen and Strayer, 1996; de Weid et al., 2006; de Wied, Gispen-de Wied and van Boxtel, 2010). Specifically, those with DBD are less likely to perceive negative facial cues in others, such as anger, fear and sadness due to dysfunctional facial mimicry response (de Weid et al., 2006, 2010; Frick and Marsee, 2007). In 2006, assistant professor of youth and family Minet de Weid led a study to investigate facial mimicry response to angry and happy expressions in boys of between 8 and 12 years of age with DBD. The results revealed that DBD boys were less facially responsive to videos of a human portraying the facial expression anger when compared to boys of the same age group, not diagnosed with DBD. EMG activity was measured in boys from both groups in the facial muscle that creates a smile action (the zygomaticus major) and the facial muscle that lowers the brow and causes a frown action (the corrugator supercilii). The boys without DBD showed both an increased corrugator muscle (frown) activity in response to angry expressions and increased zygomaticus major (smile) activity in response to happy expressions. However, while boys with DBD showed an increased smile response to happiness, they showed a significantly lower frown response to the angry expression than the boys without DBD. Due to this reduced (or lack) of facial mimicry response among DBD boys to anger, they may have been less able to understand and share this more negative emotion with others. Building on this work, in 2009 de Weid led another EMG study to investigate DBD boys' facial mimicry response to sadness and happiness evoking videos. The results revealed that DBD boys showed less corrugator facial muscle activity and a lowering of the brows to the sad video that depicted a sad scene of a boy crying over his father's absence, when compared with boys without DBD. Yet (as found in de Weid et al., 2006) no significant differences were found between the control and DBD groups for zygomaticus EMG (smile) activity on viewing more positive, happiness evoking videos.

Collectively, this infers that a lack of empathy among DBD boys applies only to negative emotions such as anger, fear and sadness and less so with happiness (see also Blair, 2006; Blair et al., 2001; Jones et al., 2009; Marsh et al., 2008). As a possible explanation as to why there is a lack of facial mimicry and empathic response in those with DBD (and wider antisocial personality disorders) it may be due to dysfunction in particular neural circuits including the amygdala (Blair, 2006, 2008; Blair and Coles, 2000; de Wied et al., 2010; Marsh et al., 2008). The amygdala is involved in the processing of visual cues of more negative emotions, especially so for fear, and it helps to alert us to potentially threatening behavior in others (Olsson and Phelps, 2007). Brain scans of DBD boys and adolescents between 10 and 17 years of age taken using fMRI revealed reduced amygdala activity when they viewed fearful facial expressions compared to a control group without DBD (Marsh et al., 2008). Without the amygdala and facial mimicry response to process and understand negative emotions such as fear, this prevents affective social interaction (de Wied et al., 2006, 2010; Marsh et al., 2008). Therefore, DBD boys are unable to understand distress-based cues and share negative emotions such as fear and sadness in others. This lack of emotional involvement and a failure to recognize more negative emotions in others can lead to a lack of control in anger and antisocial behavior as the perceived impact upon others is not recognized or as profound (de Wied et al., 2010). Based on these findings that show that DBD boys are impaired in empathy with negative emotions such as anger, fear and sadness (but not so for happiness), I suggest that there may be a lack of expectation of an empathetic response from a human-like character in DBD boys. Uncanniness is reduced for DBD boys as they themselves cannot empathize with others, especially to more negative emotions. Therefore, it does not matter if social distress cues and facial mimicry response is missing in uncanny human-like characters as DBD boys may fail to recognize this nonverbal communication. Even though this may be a selective impairment to negative emotions in DBD boys, more negative emotions such as fear are more prone to uncanniness in human-like virtual characters (see Tinwell et al., 2011, 2013). This is due to the necessity of being able to detect fear (and other negative emotions) in others promptly and respond accordingly, as a defense mechanism and to avoid a potential harm (Blair, 2003; Tinwell et al., 2011, 2013). Hence, the uncanny is reduced for boys with DBD as they may not regard human-like characters with abnormal facial expression, especially of negative emotions such as fear, as strange. However, it may be that DBD boys could still detect (and be even more sensitive to) perception of a false smile in a human-like virtual character. If the social cues to suggest true and spontaneous happiness, such as crow's feet wrinkles and bulges around the eye, were not visible, then DBD boys may still be confused if the character is attempting to communicate happiness or not. DBD is typically more common in boys than girls (Walsh, Pepler and Levene, 2002), yet DBD girls may also be less sensitive to uncanniness than healthy children. Future studies may investigate perception of the uncanny in human-like characters in DBD children.

DBD children can risk developing psychopathy in adulthood and maintain the callous, hostile and unresponsive personality traits that psychopathic individuals commonly possess (Frick et al., 1994; de Weid et al., 2010). Given this fundamental lack of empathy in psychopaths and an evident lack of concern for distressed-based social cues in others, I suggest that they too may not perceive abnormal facial expression or a lack of mimicry response in a human-like character as uncanny. A perception of apathy to another's distress and diminished empathy toward others are traits that exaggerate uncanniness in human-like characters (Tinwell, 2014; Tinwell et al., 2013), yet paradoxically if these same antisocial personality traits are present in humans (such as psychopaths), then this may prevent them from experiencing the uncanny in human-like characters. It may seem that the very traits that we perceive in uncanny human-like characters

such as a lack of mimicry response and empathy toward others may actually protect humans from the uncanny in the real world. Intriguingly, this
may work to the advantage of those who do not rely upon facial mimicry
for social interaction (such as those with Moebius syndrome or autism), or
those who lack empathy (such as DBD children and psychopaths), in that
they are less likely to feel uncomfortable and may be able to tolerate social
interaction with an uncanny human-like character for longer than healthy
people. In the next chapter I discuss the possible long-term consequences
of interacting with uncanny human-like characters on healthy individuals
and consider if we may ever overcome the Uncanny Valley.

REFERENCES

Ainsworth, M., Blehar, M., Waters, E. and Wall, S. (1978) *Patterns of Attachment*,
Hillsdale, NJ: Erlbaum.

Barabasz, A. (2013) "Evidence based abreactive ego state therapy for PTSD,"
American Journal of Clinical Hypnosis, vol. 56, pp. 54–65.

Barabasz, A. and Christensen, C. (2006) "Age regression: Tailored vs. scripted
inductions," *American Journal of Clinical Hypnosis*, vol. 48, pp. 251–261.

Blair, R. J. R. (2003) "Facial expressions, their communicatory functions and
neurocognitive substrates," *Philosophical Transactions of the Royal Society*,
vol. 358, pp. 561–572.

Blair, R. J. R. (2006) "The development of psychopathy," *Journal of Child Psychology*,
Psychiatry, vol. 47, pp. 262–275.

Blair, R. J. R. (2008) "Fine cuts of empathy and the amygdala: Dissociable deficits
in psychopathy and autism," *Q. J. Exp. Psychology*, vol. 61, no. 1, pp. 157–170.

Blair, R. J. R. and Coles, M. (2000) "Expression recognition and behavioral problems in early adolescence," *Cognitive Development*, vol. 15, pp. 421–434.

Blair, R. J. R., Colledge, E., Murray, L. and Mitchell, D. G. V. (2001) "A selective
impairment in the processing of sad and fearful expressions in children with
psychopathic tendencies," *Journal of Abnormal Child Psychology*, vol. 29,
pp. 491–498.

Bogart, K. R. and Matsumoto, D. (2010a) "Living with Moebius syndrome:
Adjustment, social competence, and satisfaction with life," *Cleft Palate–
Craniofacial Journal*, vol. 47, pp. 134–142.

Bogart, K. R. and Matsumoto, D. (2010b) "Facial mimicry is not necessary to recognize emotion: Facial expression recognition by people with Moebius syndrome," *Social Neuroscience*, vol. 5, no. 2, pp. 241–251.

Bowlby, J. (1951) "Maternal care and mental health," Report for the World Health
Organization.

Bowlby, J. (1969) *Attachment and Loss*, 3 vols. London: Hogarth Press.

Bowlby, J. and Robertson, J. (1952) "A two-year-old goes to hospital," in *Proceedings
of the Royal Society of Medicine*, vol. 46, pp. 425–427.

Bretherton, I. (1988) "Open communication and internal working models: Their role in the development of attachment relationships," *Nebraska Symposium on Motivation*, vol. 36, pp. 57–113.

Briegel, W. (2006) "Neuropsychiatric findings of Mobius sequence: A review," *Clinical Genetics*, vol. 70, no. 2, pp. 91–97.

Christensen, C., Barabasz, A. and Barabasz, M. (2009) "Effects of an affect bridge for age regression," *International Journal of Clinical and Experimental Hypnosis*, vol. 57, pp. 402–418.

Cohen, D. and Strayer, J., (1996) "Empathy in conduct-disordered and comparison youth," *Development Psychology*, vol. 32, pp. 988–998.

Corradini, A., Mehta, M. and Robering, K. (2009) "Conversational characters that support interactive play and learning for children," in Ahmed S. and Karsiti, M. (eds.), *Multiagent Systems*, Croatia: I-Tech Education and Publishing, pp. 349–374.

Di Pellegrino, G., Fadiga, L., Fogassi, L., Gallese, V. and Rizzolatti, G. (1992) "Understanding motor events: A neurophysiological study," *Experimental Brain Research*, vol. 91, pp. 176–180.

de Wied, M., van Boxtel, A., Zaalberg, R., Goudena, P. P. and Matthys, W. (2006) "Facial EMG responses to dynamic emotional facial expressions in boys with disruptive behavior disorders," *Journal of Psychiatric Research*, vol. 40, pp. 112–121.

de Wied, M., Gispen-de Wied, C. and van Boxtel, A. (2010) "Empathy dysfunction in children and adolescents with disruptive behavior disorders," *European Journal of Pharmacolagy*, vol. 626, pp. 97–103.

de Wied, M., van Boxtel, A., Posthumus, J.A., Goudena, P. P. and Matthys, W. (2009) "Facial EMG and heart rate responses to emotion-inducing film clips in boys with disruptive behavior disorders," *Psychophysiology*, vol. 46, pp. 996–1004.

Dyson, D. W. (1998). *The City of God against the Pagans*. New York: Cambridge University Press.

Fonagy, P., Steele, M. and Steele, H. (1991) "Intergenerational patterns of attachment: Maternal representations during pregnancy and subsequent infant–mother attachments," *Child Development*, vol. 62, no. 5, pp. 891–905.

Frederick, C. (2013) "The center core ego state therapy and other hypnotically facilitated psychotherapies," *American Journal of Clinical Hypnosis*, vol. 56, pp. 39–53.

Freud, S. (1919) "The uncanny," in Strachey, J. (ed. and trans.), *The Standard Edition of the Complete Psychological Works of Sigmund Freud*, (1956–1974), London: Hogarth Press, vol. 17, pp. 217–256.

Frick, P. J. and Marsee, M. A. (2007). "Psychopathy and developmental pathways to antisocial behavior in youth," in Patrick, C. J. (ed.), *Handbook of Psychopathy*, New York: Guilford Press, pp. 353–374.

Frick, P. J., O'Brien, B. S., Wooton, J. M. and McBurnett, K. (1994) "Psychopathy and conduct problems in children," *Journal of Abnormal Psychology*, vol. 103, pp. 700–707.

Gallese, V., Fadiga, L., Fogassi, L. and Rizzolatti, G. (1996) "Action recognition in the premotor cortex," *Brain*, vol. 119, no. 2, pp. 593–609.

Hatfield, E., Rapson, R. L. and Le, Y. L. (2009) "Primitive emotional contagion: Recent research," in Decety, J. and Ickes, W. (eds.) *The social neuroscience of empathy*, Boston, MA: MIT Press, pp. 19–30.

Hazan, C. and Shaver, P. R. (1994) "Attachment as an organizational framework for research on close relationships," *Psychological Inquiry*, vol. 5, pp. 1–22.

Hesse, E. (1996) "Discourse, memory and the adult attachment interview: A note with emphasis on the emerging cannot classify category," *Infant Mental Health Journal*, vol. 17, no. 1, pp. 4–11.

Holmes, J. (2001) *The Search for the Secure Base: Attachment Theory and Psychotherapy*, London: Routledge.

Ishiguro, H. (2005) "Android science: Toward a new cross-disciplinary framework," in *Proceedings of the CogSci-2005 Workshop Toward Social Mechanisms of Android Science*, Stresa, Italy, pp. 1–6.

Ishiguro, H. (2006) "Android science: Conscious and subconscious recognition," *Connection Science*, vol. 18, pp. 319–32.

James, W. (1890a) "The self and its selves," in Lemert, C. (ed.), *Social Theory: The Multicultural Readings* (2010), Philadelphia: Westview Press, pp. 161–166.

James, W. (1890b) *The Principles of Psychology*, 2 vols. Cambridge, MA: Harvard University Press.

Jones, A. P., Laurens, K. R., Herba, C. M., Barker, G. J. and Viding, E. (2009) "Amygdala hypoactivity to fearful faces in boys with conduct problems and callous–unemotional traits," *American Journal of Psychiatry*, vol. 166, pp. 95–102.

Lorenz, K. (1935) "Der Kumpan in der Umwelt des Vogels. Der Artgenosse als auslösendes Moment sozialer Verhaltensweisen," *Journal für Ornithologie*, vol. 83, pp. 137–215.

Lorenz, K. Z. (1952) *King Solomon's Ring*, New York: Crowell.

Lundqvist, L. O. (1995) "Facial EMG reactions to facial expressions: A case of facial emotional contagion?," *Scandinavian Journal of Psychology*, vol. 36, no. 2, pp. 130–141.

MacDorman, K. F. and Ishiguro, H. (2006), "The uncanny advantage of using androids in cognitive and social science research," *Interaction Studies*, vol. 7, no. 3, pp. 297–337.

Marsh, A. A., Finger, E. C., Mitchell, D. G. V., Reid, M. E., Sims, C., Kosson, D. S., et al. (2008) "Reduced amygdala response to fearful expressions in children and adolescents with callous–unemotional traits and disruptive behavior disorders," *American Journal of Psychiatry*, vol. 165, pp. 712–720.

McIntosh, D. N., Reichmann-Decker, A., Winkielman, P. and Wilbarger, J. L. (2006) "When the social mirror breaks: Deficits in automatic, but not voluntary, mimicry of emotional facial expressions in autism," *Developmental Science*, vol. 9, no. 3, pp. 295–302.

Mori, M. (2012) "The uncanny valley" (MacDorman K. F. and Kageki, N., trans.), *IEEE Robotics and Automation*, vol. 19, no. 2, pp. 98–100. (Original work published in 1970.)

Moser, M. B. and Moser, E. (1999) "Functional differentiation in the hippocampus," *Hippocampus*, vol. 8, no. 6, pp. 608–619.

Olsson, A. and Phelps, E. A., (2007) "Social learning of fear," *Nature Neuroscience*, vol. 10, no. 9, pp. 1095–1102.

Pelham, B. W., Carvallo, M. and Jones, J. T. (2005) "Implicit egotism," *Current Directions in Psychological Science*, vol. 14, pp. 106–110.

Pollick, F. E. (2010) "In search of the uncanny valley," *Lecture Notes of the Institute for Computer Sciences. Social Informatics and Telecommunications Engineering*, vol. 40, no. 4, pp. 69–78.

Prior, V. and Glaser, D. (2006) *Understanding Attachment and Attachment Disorders: Theory, Evidence and Practice*. London: Jessica Kingsley Publishers.

Rao, R. P. and Ballard, D. H. (1999) "Predictive coding in the visual cortex: A functional interpretation of some extra-classical receptive-field effects," *Nature Neuroscience*, vol. 2, pp. 79–87.

Saygin, A. P., Chaminade, T., Ishiguro, H., Driver, J. and Frith, C. (2012) "The thing that should not be: Predictive coding and the uncanny valley in perceiving human and humanoid robot actions," *Social Cognitive Affective Neuroscience*, vol. 7, no. 4, pp. 413–422.

Squire L. R. (1992) "Memory and the hippocampus: A synthesis from findings with rats, monkeys and humans," *Psychology Review*, vol. 99, pp. 195–231.

Swann, W. B., Jr. (1983) "Self-verification: Bringing social reality into harmony with the self," in Suls, J. and Greenwald, A. G. (eds.), *Psychological Perspectives on the Self*, Hillsdale, NJ: Erlbaum, vol. 2, pp. 33–66.

Swann, W. B., Jr. and Bosson, J. (2010) "Self and identity," in Fiske, S. T., Gilbert, D. T. and Lindzey, G. (eds.), *Handbook of Social Psychology* (5th ed.), New York: McGraw-Hill, pp. 589–628.

Swann, W. B., Jr. and Pelham, B. W. (2002) "Who wants out when the going gets good? Psychological investment and preference for self-verifying college roommates," *Journal of Self and Identity*, vol. 1, pp. 219–233.

Tinwell, A. (2014) "Applying psychological plausibility to the uncanny valley phenomenon," in Grimshaw, M. (ed.), *Oxford Handbook of Virtuality*, Oxford: Oxford University Press, pp. 173–186.

Tinwell, A., Abdel Nabi, D. and Charlton, J. (2013) "Perception of psychopathy and the uncanny valley in virtual characters," *Computers in Human Behavior*, vol. 29, no. 4, pp. 1617–1625.

Tinwell, A., Grimshaw, M., Williams, A. and Abdel Nabi, D. (2011) "Facial expression of emotion and perception of the uncanny valley in virtual characters," *Computers in Human Behavior*, vol. 27, no. 2, pp. 741–749.

Tronick, E. (2013) *University of Massachusetts, Boston*. Retrieved April 8, 2014, from http://www.umb.edu/research/outstanding_faculty/psychologist_ed_tronick.

Tronick, E., Adamson, L., Als, H. and Brazelton, T. B. (1975) "Infant emotions in normal and perturbated interactions," presentation at the biennial meeting of the Society for Research in Child Development, Denver, CO, April.

Turing, A. M. (1950) "Computing machinery and intelligence," *Mind*, vol. 54, no. 236, pp. 433–460.

UMassBoston (2009) "Still face experiment: Dr. Edward Tronick." Retrieved April 9, 2014, from http://www.youtube.com/watch?v=apzXGEbZht0&list=TLV6NrsJ9oA9WuuaeF0Jm2zPzR-yFB0pcA.

van der Kolk, B. (1994) "Psychobiology of posttraumatic stress disorder," in Panksepp, J. (ed.), *Textbook of Biological Psychiatry*, New York: Wiley-Liss, pp. 319–344.

Vinayagamoorthy, V., Steed, A. and Slater, M. (2005) "Building characters: Lessons drawn from virtual environments," in *Proceedings of Toward Social Mechanisms of Android Science, COGSCI 2005*, Stresa, Italy, pp. 119–126.

Walsh, M., Pepler, D. and Levene, K. (2002) "A model intervention for girls with disruptive behavior problems: The earlscourt girls connection," *Canadian Journal of Child Counselling*, vol. 36, no. 4, pp. 297–311.

Winerman, L. (2005) "The mind's mirror," *Monitor on Psychology*, vol. 36, no. 9, p. 48.

Wood, E. R., Dudchenko, P. A. and Eichenbaum, H. (1999) "The global record of memory in hippocampal neuronal activity," *Nature*, vol. 397, pp. 613–616.

CHAPTER 8

Will We Ever Overcome the Uncanny Valley?

A WARE OF THE NEGATIVE IMPLICATIONS of an empathetic, human-like virtual character falling into the valley, one of the most pertinent questions in the games and animation industries is, "Will we ever overcome the Uncanny Valley?" There is much controversy around this debate, with some announcing that the Uncanny Valley has already been crossed (Plantec, 2008), while others argue we still have a long way to go (Brodkin, 2014). In this chapter I consider the arguments for and against this question, looking to my own empirical studies that propose an alternative notion to the steep climb out of the valley. Rather than a scramble out of the valley, it may be an unsurpassable wall, unbound in height and stance (Tinwell and Grimshaw, 2009; Tinwell, Grimshaw and Williams, 2010). Based on current research and speculation, it seems that the race to overcome the Uncanny Valley is reliant on a significant shift in technology to simulate authentic realism and in our perceptual coding of nonbiological, human-like objects. It may be that one day, advances in technology and artificial intelligence will permit a human-like character to portray authentic, believable human-like behavior in real time as if one was in the company of a fellow human (Burke, 2011; Kain, 2011; Perry, 2014; Plantec, 2007, 2008). Neurological pathways may alter so that the boundaries between what stands for human and machine are more transient and we become more tolerant and less discerning of abnormal traits in artificial nonhuman-like characters (Pollick, 2010; Saygin et al., 2012).

However, I propose that neither technology nor neurological adaptation will offer a lifeline out of the Uncanny Valley. Instead, I propose that we may become less tolerant and ever more discerning of abnormalities in a human-like character's appearance and behavior as we continue to remain one step ahead of technology. As well as acquiring the technical expertise to prevent uncanniness in empathetic human-like characters, those in the games and animation industries are mindful of the financial implications of a protagonist human-like character evoking the uncanny (Freedman, 2012; Young, 2011). Furthermore, there is growing concern of the possible long-term negative health implications of continued interaction with uncanny, human-like characters for children (Perry, 2014). So here, in an attempt to reason and rationalize these concerns, I discuss what the future holds—not only how humans may react to synthetic human-like characters but also whether human-like characters may ever perceive us as odd.

8.1 OVERCOMING THE UNCANNY: A QUESTION OF TIME?

Some researchers and practitioners suggest that overcoming the Uncanny Valley is simply a matter of time and that we will become more accepting of human-like virtual characters as we grow used to them (Brenton et al., 2005; Pollick, 2010; Saygin et al., 2012). In a paper presented at the 2005 Animated Characters Interaction Workshop at Napier University, Harry Brenton and his team proposed that reaction to uncanny human-like virtual characters is not fixed but is a dynamic, changeable response (Brenton et al., 2005). Even though we may at first be put off by a human-like character's strange behavior, over time we get used to their behavior and idiosyncrasies, so that a character is no longer perceived as creepy and unsettling (Brenton et al., 2005). Brenton and his colleagues argued that "Uncanny Valley describes an emotional reaction which may be subject to change over time. Currently 'uncanny' avatars may foster a climate of their own acceptance once we are used to looking at them" (p. 3). In this sense, animators who repeatedly work with realistic, human-like characters using 3D animation software may become less disturbed by uncanny traits in the human-like 3D models that they are working with. Similarly, those who frequently watch animations with computer-generated (CG) human-like characters, or those who regularly play video games may grow less sensitive to uncanny human-like characters over time.

Others have suggested that this effect of habituation to uncanny human-like characters may be achieved as our predefined categories of what we perceive as *human* and *machine* may not be permanently fixed

(Pollick, 2010; Saygin et al., 2012). Based on her work investigating the role of the mirror-neuron system and prediction error in response to androids, the researcher Ayse P. Saygin suggested that our brains may evolve to become more accommodating of uncanny objects. "As human-like artificial agents become more commonplace, perhaps our perceptual systems will be retuned to accommodate these new social partners" (Saygin et al., 2012, p. 420). An increased exposure to hybrid human–machine characters may result in the formation of a new category in which this combined entity can comfortably reside. We would no longer experience a state of cognitive dissonance and indecision about an uncanny object, so those negative feelings associated with the uncanny may subside (Pollick, 2010; Saygin et al., 2012).

Others project that all we need is improvements in technology for imitating humans to allow us to climb out of the Uncanny Valley. Again, time is a factor required in the development of new, more powerful technologies for simulating realism with increased capabilities in motion capture, facial scanning and rendering. Developers in academia and industry continue to push the boundaries of the current limitations in technology and rely on increasing computer power to generate human-like virtual characters in animation and games, now and in the future (Burke, 2011; Kain, 2011; Perry, 2014; Plantec, 2007, 2008; Talbot, 2014). Animation commonly takes precedence over video games in this drive for technological advancement, with those new technologies and computing power transferred or adapted for games once established in animation (Kain, 2011; Perry, 2014; Talbot, 2014). Continued efforts are being made to reduce the rendering times in both the development and completion stages of a project (Kain, 2011). Ultimately it is hoped that animators will be able to render their work in real time so that the changes that they make to a 3D model are immediately updated and represented onscreen as they would appear in the final animation (Kain, 2011). At present, animators may work with a less detailed 3D model of a human-like character at a lower resolution, without seeing full details in skin texture and how light and shadows are cast on the 3D model (Burke, 2011). The model is then rendered in high definition to achieve full detail in aspects such as folds and wrinkles in a human-like character's face, skin tone, luminosity and reflection. Jeffrey Katzenburg, the chief executive of the film and animation company DreamWorks, revealed that an experienced animator may take a week to achieve just three seconds of completed animation (Burke, 2011; Kain, 2011). This hyperbolic ratio in production time is due to the large time span required for images to render (ibid.). I suggest that, given the considerable delay in animation rendering times, it may be that even if an animator observed abnormal facial expression in the final render for a human-like character, then there may not be enough time for the animator to go back and correct that character's behavior due to production deadlines. As a possible solution to the limitations that current rendering capabilities can pose, it is intended that a collaborative project between Dreamworks and the computer-chip production company Intel will provide rendering times up to seventy times faster than current rendering times of today (ibid.). Advanced computer graphics chips and multiple processing cores should allow an animator to render footage in real time, thus making the production cycle for an animated film more efficient and effective (ibid.). This technology should also have game-changing results as gaming consoles would be able to render dozens of times faster with better graphics for their human-like characters and environment. However, what would this mean in terms of the Uncanny Valley? Would the ability for animators to render in real time and for faster computational renderings in games help eliminate this phenomenon from their work? Do the exponential leaps occurring in technology actually equate to substantial leaps for human-like characters out of the Uncanny Valley?

In 2007, author and animator Peter Plantec predicted that, based on his research and wider advances in technology, it would be just two years until a human-like virtual character would evade the Uncanny Valley. "Let's hope we all get it right in the near future and move way beyond that nasty deep uncanny gully. The Holy Grail is a fully human looking, perhaps recognizable, virtual human, which we can all believe in without dissonance. I figure two more years with luck" (p. 1). Then, just 12 months later, in 2008 Plantec championed Image Metric's, human-like character Emily as having accomplished this feat. "I officially pronounce that Image Metrics has finally built a bridge across the Uncanny Valley and brought us to the other side. I was indeed wrong about it taking another two years and I'm happy about that" (p. 1). Plantec's revelation was due to the combined work of the facial animation technology company, Image Metrics, and research professor of computer science at the University of Southern California's (USC) Institute for Creative Technologies, Paul Debevec. Together, Image Metrics and Debevec pioneered the technology behind Emily (Plantec, 2008; Talbot, 2014). The basic tenets behind the CG Emily comprise markerless facial tracking techniques and high-resolution skin texture maps captured from the performance of a British female actress in her early twenties named Emily (ibid.). Emily's facial actions were first captured by Image Metrics "on a pixel-by-pixel basis" (Plantec, 2008, p. 1) defunct of facial markers. Emily's face was then scanned using a specialist light and camera rig developed by Debevec and his team at USC, so powerful it could detect individual pores on Emily's skin. The 3D scans allowed for intricate details to be captured in Emily's face including the shapes and contortions created by her underlying facial muscles and tissue and how her outer skin layer adapted and molded to these movements (Plantec, 2008; Talbot, 2014). To prevent light from scattering in many directions along different planes, the lights used in this high-definition (HD) scanning process were manipulated so that the emitted light source was focused in one direction, along one plane (Plantec, 2008). Multiple polarized lights were flashed in sequential order that highlighted every angle of Emily's face while cameras, fitted with filters to allow for this polarization, recorded the HD data (Plantec, 2008). This process created complex, consistent reference maps of Emily's face and skin texture that could be applied to the dynamic CG version of Emily (Image Metrics, 2008). It was intended that this detailed scanning process would help capture the natural variations in real Emily's different expressions based on aspects such as the strength of her facial muscles, the elasticity of her skin, her facial proportion, symmetry and how her personality may influence strength of emotion that she portrayed. In this way they could build a "perfect" (Plantec, 2008, p. 1) 3D model of Emily's face so that when the CG Emily changed facial expression, the virtual changes in her face shape, skin and underlying tissue fully replicated that of the real Emily (Plantec, 2008).

Despite a seemingly positive outlook that the Uncanny Valley would eventually be (or had already been) overcome, the findings from my experiments made me skeptical of this. The state-of-the-art CG Emily that was released in 2008 was a much improved simulacra of a human-like character since Tom Hank's CG reproduction as the Conductor in *The Polar Express* (Zemeckis, 2004) just four years prior; however, the results from my experiments showed that CG Emily, a character acclaimed as having crossed the Uncanny Valley (Plantec, 2008), was still rated below a human for perceived uncanniness (Tinwell and Grimshaw, 2009). Hence, I questioned whether it would ever be possible to overcome the Uncanny Valley. In the next section I describe the findings from my experiment to test if Emily had overcome the Uncanny Valley and if the effect of habituation, (i.e. getting used to a character) reduced sensitivity to uncanniness.

I also introduce an alternative theory to Mori's original concept of a fixed valley crevasse: a dynamic, unscalable *uncanny wall*.

8.2 THE UNCANNY WALL

Intrigued at the media response to CG Emily, this new, pioneering character was included in an experiment that I conducted in 2009 to test people's sensitivity to the uncanny in human-like virtual characters. I wanted to explore how believable and authentic CG Emily (Image Metrics, 2008) appeared when compared with a video of a human and a selection of less realistic, humanoid characters, nonhuman-like anthropomorphic-type characters and zombie characters.* Data were collected on how strange or familiar and how human-like or nonhuman-like participants judged each character to be (Tinwell and Grimshaw, 2009). To test the effect of habituation and sensitivity to the uncanny, participants were also required to disclose if they had either no experience of playing video games or using 3D modeling software, or if they had a basic or advanced experience of these activities (Tinwell and Grimshaw, 2009). Would Emily be able to match or surpass the video of the human for perceived familiarity and human-likeness to provide empirical evidence that she had overcome the Uncanny Valley? Furthermore, would extensive previous exposure to human-like virtual characters dampen a participant's sensitivity to the uncanny?

The results showed that even though CG Emily (Image Metrics, 2008) achieved higher ratings for perceived human-likeness and familiarity than all other virtual characters featured in the study, the video of the human still achieved a higher average score than Emily for human-likeness and familiarity (Tinwell and Grimshaw, 2009). In other words, CG Emily had taken second place, behind the video of the human, suggesting that she had not yet escaped the Uncanny Valley. To bridge the valley would infer that Emily was judged at least to be as familiar and human-like as the human, if not more so. Yet the results did not find this. Instead, the video of a human was still placed at the top of the valley, with human-like virtual characters unable to match or overtake the human (Tinwell and Grimshaw, 2009). Even though Plantec (2008) had previously pronounced that CG Emily had finally traversed the valley, these empirical findings showed that this innovative, of-the-moment human-like character had not yet accomplished that feat (Tinwell and Grimshaw, 2009; Tinwell et al.,

---

* Please see Tinwell and Grimshaw (2009) for a full description of the experiment design, methodology and results.

2011). Analysis of the data also revealed that participants with greater levels of experience of playing video games or using 3D modeling software were not (significantly) more accepting of human-like virtual characters than those with less or no experience of these activities (Tinwell and Grimshaw, 2009). It seemed that participants who may have had more previous interaction with human-like virtual characters in 3D modeling software or playing video games were no more willing to accommodate realistic human-like characters, such as CG Emily, than participants with less frequent and prolonged exposure to human-like characters (Tinwell and Grimshaw, 2009). This finding went against the theories of other authors (see, e.g., Brenton et al., 2005; Pollick, 2010; Saygin et al., 2012) who had speculated that human-like characters may be regarded as more humane, familiar and less strange (i.e., less uncanny) over time. Rather than increased exposure to human-like characters providing consonance, or "rightness" in the viewer's mind, this greater experience did not equate to increased acceptance (Tinwell and Grimshaw, 2009).

So what was to be made of these outcomes? Despite a considerable amount of time, money and expertise involved in creating CG Emily (Plantec, 2008), empirical evidence had revealed that she had not yet fully convinced participants that she was a human (Tinwell and Grimshaw, 2009). CG Emily's appearance and behavioral fidelity still fell short of that observed in the human video. Based on this, and that sensitivity to the uncanny may not be reduced with increased experience of interacting with realistic, human-like characters, I began to question if bridging the Uncanny Valley was (or would ever be) achievable? Also, did Mori's (1970/2012) model of an Uncanny Valley fully illustrate this perceptual problem and design dilemma? Mori's original Uncanny Valley did demonstrate the relationship between perceived human-likeness and one's affinity toward an object, yet it did not include time as a factor (Tinwell and Grimshaw, 2009; Tinwell et al., 2011). A model for human perception of synthetic human-like characters was required that accounted for time along with technological innovation.

Those designing and researching human response to androids have proposed that people may regard androids more favorably and uncanniness would potentially be reduced, if people had seen androids before and if they spent more time interacting with an android (Minato et al., 2004). However, I put forward that with respect to the uncanny, time and experience are not on our side and do not increase acceptance, but work against attempts to surmount the valley. If, on first contact, a viewer experienced the uncanny in a human-like character, then it is doubtful if more customary interaction with that character would enhance perceived believability and authenticity for that character to eventually match or exceed that of a human (Tinwell and Grimshaw, 2009; Tinwell et al., 2011). One's familiarity may increase on an interim basis as one became acquainted with that character's odd behavior, but this would not be sufficient to completely remove uncanniness. A character's flaws that a viewer may have grown accustomed to would be exaggerated again pending the application of the next breakthrough in technological realism to a novel, realistic human-like character (ibid.). This new, of the moment character would prompt the viewer to recognize all that is wrong with the earlier character, so perceived familiarity and acceptance for that earlier character are diminished (ibid.). Time is significant in our perception of the uncanny in human-like characters, but not in as much that uncanniness decreases. Instead we become ever more discerning of imperfections in a character's behavior from the human norm and increasingly sensitive to the uncanny in human-like virtual characters over time. The Uncanny Valley may be a useful concept for designers and provide a challenge to keep improving human-like characters so that they may one day reach the valley summit. However, I propose that ultimately, the answer to the question, "Can the Uncanny Valley be crossed?" is no (ibid.). Rather than a valley, we are faced with an uncanny wall that continues to grow in height. As time allows progress in technology to create human-like characters with an increased sophistication in graphical and behavioral fidelity, at the same time our perceptual expertise for detecting imperfections or abnormal behavior in a human-like character improves. Thus, the Uncanny Wall increases in height as a viewer's ability to detect imperfections from normal human behavior keeps pace with technological developments in creating high-fidelity human-like characters. Based on this analogy, the Uncanny Wall was defined as "technological discernment on the part of the audience generally keeps pace with technological developments used in the attempt to create realistic human-like characters such that, ultimately, the perception of uncanniness for such characters is inevitable" (Tinwell and Grimshaw, 2009, p. 72). A viewer's expectation of how a human-like character should behave (and respond to them) will out run technological advancements in simulating realism. In this way, we stay one step ahead of technology and prevent an artificial human-like character from "out-witting" us. Hence, state of the art characters (such as CG Emily) that may initially impress the audience are preordained to evoke the uncanny (Tinwell and Grimshaw, 2009; Tinwell et al., 2011).

While strategic design manipulations may be carried out on aspects such as a character's facial expression and speech (as a means to circumvent or enhance the uncanny in character design), the profoundly disturbing, negative response that may occur with a perceived lack of empathy in a character, alongside a viewer's intuitive ability to discern imperfections in virtual realism, suggest that we may never perceive the behavioral fidelity of a human-like, virtual character as equal to that of a human. The threat of erosion of self (identity and ego threat) is a highly uncomfortable and destabilizing experience, effectively an assault on one's peace of mind (Tinwell, 2014). As our cognitive and emotive skills increase, we may become even more aware of the premise that one cannot form an effective attachment with a human-like character or ignite a level of empathic response in that character reminiscent of others in the real world, without a subordinate threat to one's own perception of self and one's existence (Tinwell, 2014). Based on this highly unpleasant and disturbing reaction on interaction with a socioemotionally limited, human-like character that remains one step behind our expectations and perceptual expertise, overcoming the uncanny in realistic, human-like characters may be unattainable. We may continue to scale the Uncanny Wall as new human-like characters are introduced in games and animation, but there will never be the opportunity to peak the extending wall. Instead we are left looking upward at the escalating obstruction, trying to fathom how an uncanny human-like character may eventually take hold of pole position at the top of the wall.

Other researchers have suggested that, over time, humans may eventually develop new cognitive categories in which uncanny characters that represent a hybrid state between human and machine can be placed, so that they are no longer perceived as uncanny (Pollick, 2010; Saygin et al., 2012). However, I put forward that over time, as the realism for human-like characters increases (with advances in technology for simulating realism), so do our expectations of how that character will behave and respond to us. We may expect a greater level of sophistication in facial expression, such as a facial mimicry response toward us, from that character due to their increased human-likeness and a failure to provide such facial feedback evokes the uncanny. Humans may well be at a phase in ontological development whereby our perception and classification of apparently human-like, realistic virtual characters is a fuzzy concept, impairing our ability to understand and predict the future actions of such characters. As such, despite improvements in the behavioral fidelity of virtual characters, given our sophisticated social intelligence (compared with other primates and animals), this fuzziness may not clear and characters that are intended to be perceived as realistically human-like will still violate our perceptual expectations. As we continually evolve, constant detection of imperfections in realistic, human-like characters such as aberrant facial expression may only strengthen our preordained cognitive categories, not weaken their affirmed boundaries as to which objects may be classified as human or not. In this way, the foundations of our predefined boundaries are reinforced, not weakened, and no new cognitive categories are created.

This notion of an uncanny wall, ever growing in height with increased viewer discernment, may be supported in recent comments made about a novel, male human-like virtual character named Digital Ira (Activision and USC ICT, 2012). Since Emily was released in 2008, Debevec has continued to refine his 3D scanning techniques. Debevec has developed Digital Ira, an elaborative CG model headshot of a man with a shaved head, the facial features and appearance of which depict a 43-year-old actor (Perry, 2014; Talbot, 2014). Painstaking detail has been applied to ensure that aspects such as skin texture, pupil dilation and the reflective qualities of different eye components are portrayed (Talbot, 2014). In an interview with *New York Times* journalist Margaret Talbot (2014), Debevec stated that he had spent two years scrutinizing how his hyperrealistic scanning techniques would ensure an accurate depiction of the original actor. "Ira's lashes weren't quite thick enough … and the pink bit of flesh in the corner of the eye - didn't glisten effectively" (p. 32). Digital Ira can emote beyond the basic expressions of anger, disgust, fear, happiness, sadness and surprise to present facial expressions of concern, guilt and boredom (Perry, 2014). To achieve this high level of graphical fidelity and animation, NVIDIA graphics cards drive Digital Ira with nearly five trillion mathematical operations computed per second (similar to an Xbox 360 working at over forty times its maximum capacity) (Perry, 2014) that requires two teraflops of processing power (Talbot, 2014). Despite this communion of skill, expertise and dedication with the foremost capture and render technology and compute power, there have been some reports that Digital Ira may still not fully convince the viewer that he is human and Digital Ira still evokes the uncanny. In her article "Digital Actors Go Beyond the Uncanny Valley," Tekla Perry commented that she found aspects of Digital Ira's facial expression strange and unsettlingly. "Truth be told, Digital Ira is still slightly weird: his eyes often seem to stare fixedly away from the viewer, and his mouth is sometimes a little odd." In an online forum discussing the launch of a promotional video for Digital Ira, in which he speaks and emotes a smile and other facial expressions, there was a mixed response to this new human-like character. Many were amazed at the graphical fidelity and detail in Digital Ira but were not so comfortable with his speech and facial expression. In the first thread, one viewer posted, "They say they're out of uncanny valley but if you ask me, while looking incredible, it still looks kinda uncanny, especially becaus [sic] the pink skin inside your eyes, around your eyeballs is missing" (Nvidia crosses "Uncanny Valley": Digital Ira loves the Titan, 2013, p. 1). This comment shows how probing viewers are becoming to the minor details in a human-like character's appearance, as far down to the eye caruncle that Debevec had remarked and labored upon in the creation of Digital Ira (Talbot, 2014). Forum members were unhappy with Digital Ira's mouth movements and speech, as one commented, "The mouth movements are way off for me," while another posted "the Talking looks creepy tho" (Nvidia crosses …, 2013, p. 1). Aspects such as mouth articulation and lip-synchronization with speech still appeared to elicit the uncanny, as another forum member posted: "… and the mouth didnt [sic] move enough when talking" (ibid.). References were made in further comments to Digital Ira being under the influence of a drug, as if not in a normal, sane frame of mind. This sense of potential mental dysfunction in Digital Ira due to his abnormal and incongruent mouth movements emanates the connotations made between possible mental and bodily dysfunction, such as epileptic seizures, madness or the possibility of a sinister supernatural force at work behind a human-like virtual character, that I have made in previous studies investigating why asynchrony of speech and overarticulation exaggerated the uncanny in human-like characters (see, e.g., Tinwell, Grimshaw and Williams, 2010; Tinwell, Grimshaw, and Abdel-Nabi, in press). Interestingly, some found Digital Ira's smile strange and unnerving, as if it may not be a true and genuine smile with comments such as, "They still cant [sic] get him to smile convincingly," and, "That goddamn smile …" (Nvidia crosses …, 2013, p. 1). Again, these comments lend support for my previous work in human-like virtual characters (see, e.g., Tinwell et al., 2011) that perception of a false or disingenuous smile may have evoked the uncanny in Digital Ira. There are some crow's feet wrinkles displayed around Digital Ira's eye area as he smiles, but the muscles in his upper face remain tense, as if he is not relaxed and taken with the "happy," joyous moment that his smile presents. In this way his tense, awkward upper face suggests that Digital Ira may be feeling more negative emotions that contradict his smile expression, which may be confusing for the viewer. One forum member explicitly stated, "That's damn amazing ... and horrifying." This blunt comment confirms a viewer's initial awe at this highly realistic, human-like character but then the underlying, subordinate threat and fear that Digital Ira can instill as he begins to talk and emote. This type of appraisal starts to remind me of the initial suspense, marvel and then utter bewilderment at the uncanny Mary Smith created by the games company Quantic Dream, back in 2006. Eight years later, even with the latest technological realism and human expertise behind Digital Ira, it is as if we have come full circle, with previous negative critiques for uncanny Mary Smith similar to that of today's Digital Ira.

On watching this promotional video of Digital Ira (Activision and USC ICT, 2012), it is true that his voice and facial expressions do appear odd. At times, there is an overexaggeration of articulation in his lower face when Digital Ira speaks and the intensity of his gaze and piercing eyes can make one feel uncomfortable. Rather than being relaxed and at ease when I engage with Digital Ira, I feel tense and a little ill at ease. In earlier studies, I revealed a significant link between perceptions of psychopathic personality traits in an uncanny human-like character due to that character's unnatural and strange facial expression (Tinwell, Abdel Nabi and Charlton, 2013). In support of this, it seems that aspects of Digital Ira's gaze and facial expression may trigger association of psychopathic, antisocial traits for the viewer. There is no natural, subconscious break in Digital Ira's fixed stare to signal that he is thinking or contemplating what to do or say next. This heightened intensity in Digital Ira's hypnotic stare may have occurred due to eye manipulation as an *over*compensatory mechanism in an attempt to engage with the viewer. In his book *Without Conscience: The Disturbing World of the Psychopaths Among Us*, Dr. Robert Hare (1993) describes this chilling, unrelenting eye contact as a predatory, reptilian stare, typical in those diagnosed with antisocial personality disorders. Moreover, Hare recommended that people should avoid maintaining eye contact with an individual if their intense stare is making them feel uncomfortable. This psychopathic characteristic may be evident in Digital Ira as at times, I am forced to look away from Digital Ira due to his unremitting, predatory-like stare. This strong stare emanates a superficial charm as if Digital Ira lacks integrity and may be of an untrustworthy nature. Overall, Digital Ira's behavior appears unnatural and false, as if he is attempting to deceive the viewer that he is of a pleasant and happy demeanor, when he is actually hiding more negative emotions and selfish intentions. This behavior may betray Digital Ira as having possible antisocial psychopathic traits that make the viewer feel apprehensive and uncomfortable as one perceives Digital Ira is scheming how they may be used to his own self-gain. If Digital Ira is to put the viewer at ease, then his sleightful facial expression should be corrected. The tense muscles in Digital Ira's upper face including his forehead and brows and his piercing stare are inconsistent with his lower face that is sometimes posed in a more relaxed, neutral expression or smile. Hence, it is as if Digital Ira is constantly on alert and unable to relax and I find myself thinking, "OK, so he looks like he is resting, or is presenting a smile, but what does he really want and what is he really thinking?" Currently, Digital Ira cannot relationally interact with humans, but once technology enables Digital Ira (and other characters like him) to communicate in real time, it may be that these feelings of unease do not dissipate but intensify for the viewer.

Not all are negative, and there have been reports made that Digital Ira (Activision and USC ICT, 2012) has crossed the Uncanny Valley. Writing for *Forbes*, Jason Evangelho (2014) announced that the advent of Digital Ira signals farewell to the Uncanny Valley and that the graphical fidelity for this character is sufficient to prevent perception of the uncanny, "Digital Ira has lifelike traits without dropping off the creepy cliff of the uncanny valley" (p. 1). However, he also remarked that aspects of Digital Ira's behavioral fidelity, such as speech and emotion, remain a challenge for the professionals fostering this striking animation. "Incredible as the still shot of Ira is … the illusion is usually shattered when it comes to speech. Tying in vocalization with smoothly transitioned expressions and convincing emotions is no easy feat" (p. 1). As with CG Emily (Image Metrics, 2008), further empirical investigation is required to assess if Digital Ira has at last overcome the Uncanny Valley and to consolidate conflicting feedback for this elaborate character. The uncanny wall supposes that our technological discernment for detecting imperfections in realistic human-like characters (especially inadequate facial expression) will keep in line with technological developments used to create such characters, so they will inevitably be regarded as uncanny. To test this theory, future longitudinal experiments will be required to establish if improved behavioral fidelity and facial expression in human-like characters may eradicate the uncanny, or, if they are preordained to be regarded as uncanny despite improvements in simulating realism. The former result would provide some hope that current (and future) developments in technology used to simulate realism do allow for a greater sophistication of dynamic facial expression (such as reciprocal facial mimicry) and speech in a character to allow us to accept them as human: the latter providing validation for the potential barrier that an uncanny wall poses. The latest next-generation, realistic, human-like character may be compared to a video of a human and earlier human-like characters to help establish if characters may be perceived as more or less uncanny over time and if viewer perception of new, pioneering characters may ever be comparable to a human or even overtake human footage. For example, a similar experiment to that conducted in 2009 with CG Emily (Tinwell and Grimshaw, 2009) may be rerun that includes, among anthropomorphic, zombie, stylized human-like and other realistic human-like characters, CG Emily (created in 2008), a video of a human and the innovative Digital Ira. I predict that not only will CG Emily be rated as stranger and less human-like than a human (i.e., more uncanny), but the difference between a human and CG Emily will have increased further than the scores achieved in the first experiment conducted in 2009 (Tinwell and Grimshaw, 2009). This gap between CG Emily and the human for uncanniness ratings will have increased due to enhanced audience discernment as they may find CG Emily increasingly primitive, naive and strange over time. Furthermore, Digital Ira may be perceived as less uncanny than CG Emily and be ranked higher than CG Emily on the Uncanny Wall, but the pioneering technology behind Digital Ira may not be sufficient to level with or surpass the human video. In other words, the audience will not succumb to the latest "technical trickery that attempts to persuade of the humanness of the character" (Tinwell and Grimshaw, 2009, p. 72). I predict that in the long run new versions of Digital Ira and other new characters created using the latest emerging technologies and expertise may also suffer the same fate, ranked beneath a human and destined for uncanniness as the human simulacra remains out of reach.

8.3 THE HUMAN AND FINANCIAL COST OF UNCANNY HUMAN-LIKE CHARACTERS

As well as their wide usage in games and animation, realistic human-like characters are increasingly being used in therapeutic applications to help counsel and educate patients, to train health practitioners, to motivate individuals to exercise and to serve as companions for the elderly (Keskin et al., 2007; Nairn, 2013; Topol, 2012). Yet the possible negative human and financial implications of featuring uncanny human-like characters in digital applications suggest that we should approach with caution. If children are exposed to games and animation with uncanny human-like characters both at home and in school, then there may be a possible risk of this uncanny facial expression of emotion in a human-like character dampening how a child then relates to other people. It may be that frequent interaction with uncanny, human-like characters with aberrant facial expression may have a detrimental effect on a child's social development as they are less able to recognize and react to another's nonverbal facial cues, not only with onscreen virtual characters but in the real world. The increasing time spent with emotionally limited human-like characters may start to alter our perception of what is acceptable in human social interaction, especially in children who are still refining their perceptual expertise and social skills. Children may become less adept at recognizing subtle, nonverbal facial cues in others if this is the "norm" in human-like characters featured in the virtual worlds that they encounter. As I have anticipated, it is unlikely that characters with a realistic human-like appearance will ever be perceived as portraying an emotional sophistication and behavioral fidelity equal to that of a human. Yet, time spent interacting with uncanny human-like characters may still increase in both children and adults as such characters are continually introduced as a novel and convenient interactant in social environments. As well as in games and animation for entertainment purposes, they are likely to feature as relational customer service agents in society and be relied upon to communicate information to the public. Even if a realistic human-like character looks convincingly human, there lies a risk of a child (or adult) being exposed to a potentially ugly situation with a lack of fully reciprocal and responsive facial expression and human–character interaction. If this ugly situation (Tronick et al., 1975) is frequent and prolonged due to the necessities of acquiring vital information from these emotionally limited human-like characters, then returning to effective social interaction in the real world may be impaired. Even if this malfunction in social communication occurs as a momentary or temporary glitch in a child's conceptual expertise and social skills, the negative, harmful ramifications of this lower-level communication with a human-like character may still hinder that child's social development on a long-term basis. To paraphrase, as I stated in an interview with IEEE editor Tekla Perry (2014), "When children grow up spending much of their days interacting with realistic computer-generated characters they may risk some of their ability to understand flesh-and-blood people and be handicapped when it comes to interacting with the real thing." This aspect of possible social harm and deprivation in children on encountering uncanny realistic human-like characters may reveal itself as the darker, more sinister side of simulating realism in technology.

As well as this possible human cost, some may have suffered monetary loss by creating and featuring human-like virtual characters that were intended to be empathetic, but not perceived as so, in a video game or movie. A comparison could be made between the cinematic drama game *Heavy Rain* (Quantic Dream, 2010) and the action thriller *Grand Theft Auto V* (Rockstar North, 2013). As discussed in previous chapters of this book, empathetic, protagonist characters such as Ethan Mars were criticized for being uncanny in *Heavy Rain* (Quantic Dream, 2010). This character's unnatural facial expression and speech was perceived as odd and peculiar, to the extent that, against the intentions of the game designers, some players could not relate to and show pity toward Ethan (Hoggins, 2010). Even though the uncanny still occurred in human-like characters featured in *GTA V* (Rockstar North, 2013), such as Michael De Santa's incongruent and chaotic facial expression and articulation, uncanny facial expression and speech may have worked to the advantage of this antipathetic character and the game's overall success. As a game that focuses on the crime-laden underworld of Los Santos, the viewer may expect more unusual behavior in antipathetic characters that would suggest the possibility of hostile, antisocial traits in a character. Uncanny facial expression and speech enhanced how untrustworthy and insincere these characters were perceived to be, in keeping with what the designers had intended and player expectations for this game. As the fastest-selling game of all time and having set new sale records for any entertainment property (Lynch, 2013), in this case, the uncanny may have contributed to *GTA V*'s success. *Heavy Rain* sales still attained what Quantic Dream had predicted; however, given the global anticipation of *Heavy Rain*, this game may have achieved even better, record sales had the fundamental protagonist characters not been plagued by the uncanny. Thus, uncanniness had an adverse effect on player engagement and contradicted what the player may have expected in this game.

A comparison of estimated budgets and gross film takings infers that there may be a negative financial impact of uncanny human-like virtual characters in feature films. *Mars Needs Moms* (Wells, 2011) had an estimated budget of 150 million dollars (Internet Movie Database [IMDB],

2011; Young, 2011), and when it was released in March 2011, it achieved a "disastrous $6.9 million opening" (Young, 2011, p. 1) over its first weekend cinema debut in the United States of America (USA). This disappointing start yielded further financial failure with gross takings by June 2011 at $21,379,315 in the United States (IMDB, 2011). This uncanny animation made less profit in the four months from opening than James Cameron's acclaimed *Avatar* (Cameron, 2009). While *Avatar* had a greater budget of 237 million dollars (IMDB, 2009), the successful design and inclusion of *less* human-like CG protagonist Na'vi characters aided the overall success of this film. *Avatar* grossed a profitable US$740,440,529 between opening in December 2009 and March 2010 (IMDB, 2009). Even though the CG Na'vi character's facial expression and behavior was simulated from captured human performance, nonhuman characteristics of the Na'vi race such as blue skin and large ears reduced a viewer's expectations. Therefore, any deviations from the human norm in a Na'vi character's facial expression and behavior was more acceptable and did not elicit the uncanny. The audience may not have expected the Na'vi characters to emote happiness or fear, but when they did, the viewers could recognize and relate to these human-like emotions. A spectator could impute their own sense of virtue onto Na'vi characters and understand their feelings and intentions. In this way, Na'vi characters that were intended to be perceived as empathetic and likeable were perceived as so, as the viewer could engage with their emotional state. The viewer had already accepted that the Na'vi were not fully human due their nonhuman-like features and appearance. Yet, any signs of emotion in these characters that the viewer could relate to enhanced how likeable and effective their performance was in this film so that they were regarded positively. Cameron had generated human-like emotion and behavior in his CG Na'vi characters without causing unfulfilled expectation and uncanniness for the viewer, which contributed to this film's success. The weaker box office performance of *Mars Needs Moms* was speculated to have brought about the demise of Robert Zemeckis's motion-capture studio ImageMovers Digital (Freedman, 2012; Young, 2011). In this way, unpopular, uncanny human-like characters may provide a causal effect as to why *Mars Need Moms* was not as popular, profitable or successful as *Avatar*, which featured nonhuman-like, accepted, protagonist characters.

Attributions to the uncanny are still rife in animation and games, and, as Mark Daly, senior director of content development at the company NVIDIA, recently admitted, "I think every game out there right now has an uncanny valley problem" (Brodkin, 2014). Even Masahiro Mori, portender of the uncanny in synthetic human-like agents, has declared that he has no wish to develop android robots with a high human-like fidelity (Kageki, 2012). From Mori's perspective, it is not worth the risk of attempting to cross the Uncanny Valley as he feels that "robots should be different from human beings" (p. 1). Given this general consensus that the uncanny still remains in human-like virtual characters and that some researchers conclude that humans should remain distinct from man-made artificial entities, the questions remain, "What does the future hold?" and "Where do we go from here?" It may be the safest and arguably, the most cost effective strategy to just include characters with a reduced human-like appearance in games and animation, but would this not be an admission of defeat? Humans continually strive to improve themselves and society, which requires a comprehensive understanding of our mental and physical capabilities and how they are attuned and connected with one another. Rather than immediately projecting the "best possible" but inadequate reproduction of a human into an animation or game for commercial gains, the goal for developing human-like virtual characters should be to gain a better understanding of our own make up. The compromises that designers and developers may make in the final, rendered version of a human-like character due to comprises and limitations in technology are unacceptable and foster audience resilience and rejection. It is prudent therefore to reflect on what we know (and still don't know) about human development from birth and, importantly, how this knowledge may be transmitted to a "human-like virtual newborn." A more systematic way of working is required that will provide a site for accumulating and consolidating knowledge of the human biological and conceptual being. I duly summarize by considering how a human-like character may evolve from a virtual newborn, with learned cognition, emotion and perceptual skills that may allow the human-like character to perceive uncanny humans, thus evading uncanniness itself.

8.4 THE FUTURE: A HUMAN-LIKE VIRTUAL NEWBORN

With regards to the future, I would hope that we will one day be able to design and develop human-like relational agents with the sophistication and capacity to successfully socially interact with humans without a negative uncanny response in an individual. I hope that human-like relational agents presented either onscreen, as 3D holograms, or as androids will be able to achieve full multimodal, affective communication with us, based on their ability to successfully read and respond to our emotive state. This may be through visual, acoustic, sensory and haptic feedback mechanisms. In other words, they not only understand our thoughts and feelings, but can respond intuitively to us, for example with a sense of humor and compassion. They would be able to detect when we felt sad or melancholy and adapt their behavior to our mood. In this way, human-like relational agents may one day be used to help humans and forge meaningful relationships with us. If we can interact with human-like virtual characters in a successful and harmless way, then such companionship with a human-like character may be beneficial, from helping to prevent or reduce loneliness in an increasing aging population, to having a human-like relational agent as a valued member of a work team.

Before this can happen though, intelligence must be available to allow for smooth social interaction and the human-like agent to show humility and empathy toward others; otherwise our ability to forge an effective and meaningful attachment with such an agent may be prevented by experience of the uncanny. From a more skeptical perspective, it may be that humans will always be alerted to a perceived difference in human-like characters due to our increasing discernment that keeps pace with technology and an innate instinct to protect ourselves from possible harm. However, one day technology may be just good enough to persuade us that human-like relational agents are believably and authentically human, with which we may forge valuable attachments.

At the moment we are asking of developers and animators to develop fully convincing, authentically believable emotive human-like characters without a full understanding of the psychological and physical processes behind human emotion and cognition. For example, it is still not certain whether we make a happy facial expression and then, because the brain has interpreted this facial expression, we then process that we are feeling happy, or, if the neural pathways for experiencing happiness are triggered first, which then prompts us to present and communicate a happy facial expression (Kleinke, Peterson and Rutledge, 1998). Researchers still debate if a heightened arousal of our autonomic nervous system then triggers a facial expression of fear, or if an external fear/shock facial expression is required before our inner cognitive and nervous systems can respond to this negative emotion (Kleinke et al., 1998). Theories in emotion differ about whether physiological response to an emotion changes our facial expression, or the other way around (Kleinke et al., 1998). Without a confirmatory, precise understanding of human psychology and physiology, man-made human-like characters may present this ambiguity in our understanding of human behavior so that they portray abnormal maladaptive emotional behavior. As Joe Letteri at Weta Digital studio proclaims, emotive human-like characters need to be generated "from the inside out, to be able to ask, what is this character thinking, and how does that relate to his face?" (Perry, 2014). I propose that the secret may lie in eventually overcoming the uncanny, when a synthetic human-like character has the innate experience itself to recognize when a human has maladaptive and abnormal social skills. For example, a human-like character would be able to detect the uncanny in us if it was presented with a human who could not mimic and understand its facial expressions and who lacked the propensity to engage with that character. If we turned the previous methodology that I have used in experiments on its head (see Tinwell et al., 2011, 2013) and presented a human-like virtual character with a blank, expressionless human, would the character be able to detect something wrong with the human and want to reject them? In this case, rather than a human judging fully animated and partially animated human-like characters with a lack of upper facial movement for uncanniness, the human-like character would be the participant. The stimuli may be grouped as: humans with full facial movement and humans that could not move their upper facial muscles, including a lack of expressivity in the eye region, eyebrows and forehead. This lack of nonverbal communication in the upper face may be due to people having had Botox treatment to eliminate upper facial movement or being of a psychopathic nature so that they do not present a facial startle response (wide eyes and raised brows) to fearful events (Benning et al., 2005; Herpertz et al., 2001; Justus and Finn, 2007). Both human groups would present different facial expressions of emotion, but the perceptual expertise of the human-like character should be good enough to detect when there is something not quite right and possibly disturbing about the aberrant facial expression in the human group with a lack of upper facial movement. That, in essence, may be the true test of overcoming the Uncanny Valley: when a human-like virtual character can sense a lack of empathy and facial mimicry in us toward that character, to the extent that we are perceived as strange and uncanny if we behaved in such a way. But, how could this goal be achieved? Given that the triggers that alert us to uncanniness may be innate, survival tactics, such as an intrinsic ability to detect and respond to others' negative and possibly threatening behavior and the instinct to forge attachments with others, it is these very qualities that would need to be ingrained in a human-like virtual character. To date, the majority of human-like characters are created as adults, such as Emily (Image Metrics, 2008) and Digital Ira (Activision and USC, ITC, 2012), with some children and adolescent characters featured in animations such as *The Polar Express* (Zemeckis, 2004), *Tintin* (Spielberg, 2011), *Mars Needs Moms* (Wells, 2011), and the video game *Heavy Rain* (Quantic Dream, 2010). However, by that age, the human-like virtual characters should have acquired the majority of fundamental perceptual and social skills to provide smooth and effective social interaction and be alerted to a perceived strangeness in someone else with aberrant facial expression. We are born with some of these communication skills and behaviors such as forging attachments (Bowlby, 1969; Hazan and Shaver, 1994). Then, socioemotional skills are learned and further developed in the first two years of life such as interpreting and responding to others' facial expressions (Bowlby, 1951, 1969; Swann and Bosson, 2010; Tronick et al., 1975). Therefore, to overcome uncanniness in human-like virtual characters, we may need to develop them from virtual birth so to speak, as a newborn and not as an adult. In this way the virtual human-like infant has a chance to learn and equip itself with the crucial and fundamental social, emotional communication skills that it will rely on to successfully interact with others. We must be sure that at each stage of that virtual infant's development, it is learning and presenting the social skills and behavior that a human would. Intelligence would be required to allow the virtual infant to perceive when it was being ignored by others and to trigger a protest–despair–detachment reaction, to prevent bad or ugly situations from occurring that impede its developing and future social skills. In this way, the virtual infant may keep pace with how a human infant would learn and behave in its first two years. Then, as it grows into childhood, if it is lacking any important social skills this would become evident as in the real world with those diagnosed with autism and disruptive behavior disorders. By implementing sufficient intelligence to enable this real-time growth and development for the virtual human-like infant, it may gain the experience and ingrained skills to successfully interact, empathize and forge attachments with others as a human would. At the moment, when developing adult human-like virtual characters, we are working from the top downward and trying to invent past experience and communication skills that would otherwise have taken years to develop. Therefore, it is not surprising that we develop uncanny human-like characters that present antisocial personality traits, as they have not had a full opportunity to learn good socialization skills. If we developed a virtual human-like character from a virtual newborn entity, then we may have the chance for that virtual character to acquire the intelligence and aptitude to develop effective social skills, a sense of self-concept, independent thought, an individual identity and ego.

One notable previous attempt at recreating a human-like infant was in an animated short named *Tin Toy,* directed by John Lasseter in 1988. Billy, a human-like baby dressed in a nappy, attempts to chase and pester a mechanical, one-man-band toy character who tries to escape the baby's calamitous pursuit. A lack of upper facial expression exaggerated the uncanny for this baby in that while his mouth moved to show his surprise and delight at finding new toys, his upper face did not communicate these emotions. This lack of congruence between the baby's upper and lower face confused and irritated the viewer, with reports of how ugly and unrealistic viewers perceived Billy to be (Talbot, 2014; *Tin Toy* Reviews & Ratings, IMDB, 2002). Rather than being perceived as cute and inquisitive, the baby was perceived as creepy and disturbing with Machiavellian traits. Billy's peculiar facial expression and behavior prevented some from engaging and empathizing with him. "All of the toys in that movie are wonderfully animated. The baby was horrible. Its drool was like Silly Putty. It inspired no empathy" (MacDorman, as quoted in Talbot, 2014, p. 33). The baby's awkward mouth movements and creepy smile lacked the innocence and naivety that one may expect in a baby. Instead of considering what perceptual, emotional and cognitive skills a young infant should portray at that age and its innate behavior to forge attachments with others, Billy was just "delivered" on screen with attitudes and behaviors beyond a baby's years. Billy's crude facial expression was criticized for uncanniness in that it appeared older than the baby's chronological age, hence depicting abnormal and strange behaviors in Billy that went against his younger, physical appearance.

The aims and scope of creating a human-like virtual newborn character would no doubt be beyond an individual's lifetime of work and would require a team of developers and psychologists to enable the virtual infant to progress to childhood and eventually adulthood. Moreover, consistent, responsive caregivers may be required to help bring up the virtual character and allow it to learn from and bond effectively with humans. Then, this human developmental knowledge kindled with developments in virtual-realism technology may allow us to develop a convincing, true to life, human replica. If continued, successful intelligence and development can be maintained for the virtual character, on a cognitive and physical level that keeps pace with human development; then we may be closer to overcoming the uncanny in human-like characters. The human-like characters onscreen today appear as a ready-made child or adult that is not equipped with or has acquired the essential, effective social skills to aid survival in humans. Therefore, we will continue to perceive differences or strangeness in the way that a human-like virtual character thinks and behaves. The human-like character needs to understand the importance of effective facial expression in others and itself and recognize when there is aberrant facial expression in another. Such skills may only ever really, convincingly be learned and acquired from long-term progressive interaction with others, as an infant would in the real world before it can learn to interact with humans as an adult would. The cognitive and physical development of the human-like virtual infant could be checked at stages comparable to those conducted on newborns and infants in the real world, such as pupil dilation to light tests and various speech tasks. In this sense, any abnormalities in the virtual infant's development and growth could be corrected using appropriate intelligence, so that the character stays on track with humans of the same age group and ability. If at any stage issues with social skills and emotional intelligence are overlooked or are not fully developed, then the character risks maintaining these dysfunctional and/or underdeveloped skills into later life, that risk being judged as abnormal and uncanny. The modifications made to the human-like virtual character as it grew up, to correct any observed abnormalities in the character's appearance and behavior that did not match a human of a similar age, would be crucial on both a cognitive and physical level. If at any moment in the character's development, that it was judged to show antisocial traits or abnormal behavior, then the intelligence and graphical details to simulate normal social and emotional development in humans would have to be retraced and corrected to ensure it was evolving as a healthy human would. This would not only provide the human-like character an opportunity to develop and portray a personality, but its physical appearance would be authentically changed with age to suit the needs of that human-like character, not the designer behind the human-like character.

The moral and social implications behind developing a human-like virtual character from a newborn state are salient in such a project. The human-like virtual character may be modeled on a particular newborn, so that it forms a simulacrum of that particular person throughout their lifespan. However, this may present issues associated with a doppelganger effect that may negate that mirrored person and have a harmful, not helpful, impact on their life. Would it be fair and ethical to create a twin of an individual from birth without their individual consent? Even if their caregivers had provided consent, the implications for that individual would be profound and central to its own life. In retrospect, it may be better to adapt a more general approach and develop a human-like virtual character based on average statistics within healthy individuals of the population. Data may be monitored and used to guide the design and development of the human-like virtual character from studies and statistics gathered on human cognitive, perceptual, social and physical features, aptitudes and abilities from birth and at differing ages. The human-like character may recognize and "know" people that it has bonded with from earlier life, to those who are strangers who it has not previously encountered. It may be advantageous to develop several different models of this newborn virtual character who may interact with different people (and caregivers), in different languages and across different cultures to ensure that the behaviorisms and mannerisms developed by the human-like virtual character are representative of and familiar to those who will interact with it. In this respect, the human-like virtual character may fit in with its local society and be more foreign outside its learned society, as in real life. The human-like character would keep evolving in perceptual expertise and communication skills as humans evolve, so that it may stand the chance of being perceived as in affective harmony and at the same level with us, without us being one step ahead. Further moral ethical implications may still arise from such a project, as to how long a human-like virtual being should exist or if it may have an eternal lifespan. A sense of dread of one's own death and mortality have been intrinsically linked with the uncanny, so it may be that for a human-like character to encapsulate what it means to be human, it cannot represent an ethereal immortality, but it too must be aware and contemplative of its own survival needs and inevitable death.

When we first interact with a human-like virtual character such as Emily (Image Metrics, 2008) or Digital Ira (Activision and USC ICT, 2012), the character is a stranger not only to us but also to themselves. Without an understanding of its past, present and possible future states, it is as if we are interacting with someone who has dementia. They lack an integral knowledge of where they have come from and how this experience and memory helps inform their present situation and what they may do in the future. It is just an immediate, acknowledgement of the present, with no (or a limited) perception of what has previously occurred and what the future may hold, and why this information is important to them. If a human-like virtual character was designed from birth, as I have proposed above, then its continued development and growth, in keeping with human development, may provide theory of mind in that character. A systematic approach to a character's development and learning to enable evolutionary as well as developmental traits may enable a human-like character to be perceived as authentically, believably, human. The experience and expertise that it may gain as a human would, may allow the virtual character the ability to make independent decisions and hold an opinion based on its previous experience and learning. An understanding of what is right and wrong and individual preference for certain people and things may help maintain and reinforce that character's sense of self and virtual ego. Beyond comprehension of the basic emotions, the character may understand when another person is being sarcastic and have the ability to present sarcasm itself; for example, to interpret how one situation may appear funny or out of context with another, with a more humorous connotation. Or to realize when persistently staring at a person for a long period of time may be inappropriate and make that person feel uncomfortable, as that person may then take a more defensive stance toward the character. The human-like virtual character may then interact with children, adults and the elderly in a helpful, beneficial way, without the risk of a harmful, uncanny interaction. Without these fundamental abilities gained by continued, developmental intelligence and responsive interaction with others, we may continue to be presented with a shallow, hollow depiction of an uncanny human-like character, without the depth and experience to respond accordingly to others and their behavior. In summary, we currently create human-like virtual characters with the very traits that we find strange and uncanny in humans, such as abnormal facial expression and antisocial personality traits. To prevent this, we must acquire a new design and development methodology for human-like virtual characters that begins in life as a human would, ready to learn, receive and give as part of the real world.

REFERENCES

Activision Inc. and USC Institute for Creative Technologies (2012) *Digital Ira* (Facial Animation). California: Activision and USC ICT.

Benning, S. D., Patrick, C. J. and Iacono, W. G. (2005) "Fearlessness and under-arousal in psychopathy: Startle blink modulation and electrodermal reactivity in a young adult male community sample," *Psychophysiology*, vol. 42, no. 6, pp. 753–762.

Bowlby, J. (1951) "Maternal Care and Mental Health," Report for the World Health Organization.

Bowlby, J. (1969). *Attachment and Loss*, 3 vols. London: Hogarth Press.

Brenton, H., Gillies, M., Ballin, D. and Chatting, D. (2005) "The Uncanny Valley: Does it exist?" paper presented at the *HCI Group Annual Conference: Animated Characters Interaction Workshop*, Napier University, Edinburgh, September 5–9.

Brodkin, J. (2014) "Game developers crossing the Uncanny Valley." Retrieved August 20, 2014, from http://news.dice.com/2014/02/14/game-developers-crossing-uncanny-valley/.

Burke, A. (2011) "Cutting Wait Time: DreamWorks-Intel Partnership to Revolutionize Rendering." *Forbes*. Retrieved August 20, 2014, from http://news.yahoo.com/cutting-wait-time-dreamworks-intel-partnership-revolutionize-rendering-182606985.html.

Cameron, C. (producer/director) (2009) *Avatar* [Motion picture]. Los Angeles, CA: Twentieth Century Fox.

Evangelho, J. (2014) "Good bye Uncanny Valley: NVIDIA'S 'face works' brings shocking realism to facial animation," *Forbes*. Retrieved May 1, 2014, from http://www.forbes.com/sites/jasonevangelho/2013/03/20/goodbye-uncanny-valley-nvidias-face-works-brings-shocking-realism-to-facial-animation/.

Freedman, Y. (2012) "Is it real ... or is it motion capture? The battle to redefine animation in the age of digital performance," *Velvet Light Trap*, vol. 69, pp. 38–49.

Grand Theft Auto V (2013) [Computer game], Rockstar North (Developer), New York: Rockstar Games.

Hare, R. D. (1993) *Without Conscience: The Disturbing World of the Psychopaths among Us*. New York: Guilford Press.

Hazan, C. and Shaver, P. R. (1994) "Attachment as an organizational framework for research on close relationships," *Psychological Inquiry*, vol. 5, pp. 1–22.

Heavy Rain (2010) [Computer game]. Quantic Dream (Developer), Japan: Sony Computer Entertainment.

Herpertz, S. C., Werth, U., Lukas, G., Qunaibi, M., Schuerkens, A., Kunert, H. J., et al. (2001) "Emotion in criminal offenders with psychopathy and borderline personality disorder," *Archives of General Psychiatry*, vol. 58, no. 8, pp. 737–745.

Hoggins, T. (2010) *Heavy Rain* video game review," *Telegraph*. Retrieved October 31, 2013, from http://www.telegraph.co.uk/technology/video-games/7196822/Heavy-Rain-video-game-review.html.

Image Metrics (2008) *Emily Project* (Facial Animation), Santa Monica, CA: Image Metrics, Ltd.

Internet Movie Database (2009) *Avatar*. Retrieved May 11, 2014, from http://www.imdb.com/title/tt0499549/business?ref_=tt_dt_bus.

Internet Movie Database (2011) *Mars Needs Moms*. Retrieved May 11, 2014, from http://www.imdb.com/title/tt1305591/.

Justus, A. N. and Finn, P. R. (2007) 'Startle modulation in non-incarcerated men and women with psychopathic traits," *Personality and Individual Differences*, vol. 43, no. 8, pp. 2057–2071.

Kageki, N. (2012) "An uncanny mind: Masahiro Mori on the Uncanny Valley and beyond," *IEEE Spectrum*, Retrieved December 2, 2013, from http://spectrum.ieee.org/automaton/robotics/humanoids/an-uncanny-mind-masahiro-mori-on-the-uncanny-valley.

Kain, E. (2011) "Techonomy: Can Intel and DreamWorks Cross the Uncanny Valley?" *Forbes*. Retrieved August 20, 2014, from http://www.forbes.com/sites/erikkain/2011/11/14/techonomy-can-intel-and-dreamworks-cross-the-uncanny-valley/.

Keskin, C., Balci, K., Aran, O., Sankur, B. and Akarum, L. (2007) "A multimodal 3D healthcare communication system," In *3DTV Conference: The True Vision—Capture, Transmission and Display of 3D Video*, Kos, Greece, pp. 1–4.

Kleinke, C. L., Peterson, T. R. and Rutledge, T. R. (1998) "Effects of self-generated facial expressions on mood," *Journal of Personality and Social Psychology*, vol. 74, pp. 272–279.

Lasseter, J. (1988) *Tin Toy* [Animation Short] Emeryville, CA: Pixar.

Lynch, K. (2013) "Confirmed: *Grand Theft Auto 5* breaks 6 sales world records," Guinness World Records. Retrieved May 11, 2014, from http://www.guinnessworldrecords.com/news/2013/10/confirmed-grand-theft-auto-breaks-six-sales-world-records-51900/.

Nairn, G. (2013) "Meet George Jetson, M. D.," *Wall Street Journal*. Retrieved May 11, 2014, from http://online.wsj.com/news/articles/SB1000142405270 2304636404577291252691059524

Nvidia crosses "Uncanny Valley": Digital Ira loves the Titan (2013) [online forum] *Face Punch Studios*. Retrieved May 5, 2014, from http://facepunch.com/showthread.php?t=1255355

Perry, T. (2014) "Digital Actors Go Beyond the Uncanny Valley," *IEEE Spectrum Magazine*. Retrieved August 28, 2014, from http://spectrum.ieee.org/computing/software/digital-actors-go-beyond-the-uncanny-valley.

Plantec, P. (2007) "Crossing the great Uncanny Valley," *Animation World Network*. Retrieved April 29, 2014, from http://www.awn.com/articles/production/crossing-great-uncanny-valley.

Plantec, P. (2008) "The digital eye: Image Metrics attempts to leap the Uncanny Valley," *Animation World Network*. Retrieved April 29, 2014, from http://www.awn.com/vfxworld/digital-eye-image-metrics-attempts-leap-uncanny-valley.

Pollick, F. E. (2010) "In search of the uncanny valley," *Lecture Notes of the Institute for Computer Sciences. Social Informatics and Telecommunications Engineering*, vol. 40, no. 4, pp. 69–78.

Reviews & Ratings for Tin Toy (2002) *Internet Movie Database*. Retrieved May 8, 2014, from http://www.imdb.com/title/tt0096273/reviews?start=0.

Saygin, A. P., Chaminade, T., Ishiguro, H., Driver, J. and Frith, C. (2012) "The thing that should not be: Predictive coding and the uncanny valley in perceiving human and humanoid robot actions," *Social Cognitive Affective Neuroscience*, vol. 7, no. 4, pp. 413–422.

Spielberg, S. (producer/director) (2011) *The Adventures of Tintin: The Secret of the Unicorn* [Motion picture]. Los Angeles, CA: Paramount Pictures.

Swann, W. B., Jr. and Bosson, J. (2010) "Self and identity," in Fiske, S. T., Gilbert, D. T. and Lindzey, G. (eds.), *Handbook of Social Psychology*, New York: McGraw-Hill, pp. 589–628.

Talbot, M. (2014) "Pixel Perfect: The Scientist Behind the Digital Cloning of Actors," *New Yorker*. Retrieved August 29, 2014, from http://www.newyorker.com/magazine/2014/04/28/pixel-perfect-2.

Tinwell, A. (2014) "Applying psychological plausibility to the Uncanny Valley phenomenon," in Grimshaw, M. (ed.), *Oxford Handbook of Virtuality*, Oxford: Oxford University Press, pp. 173–186.

Tinwell, A., Abdel Nabi, D. and Charlton, J. (2013) "Perception of psychopathy and the Uncanny Valley in virtual characters," *Computers in Human Behavior*, vol. 29, no. 4, pp. 1617–1625.

Tinwell, A., Grimshaw, M. and Abdel Nabi, D. (in press) "The effect of onset asynchrony in audio-visual speech and the Uncanny Valley in virtual characters," *International Journal of the Digital Human*.

Tinwell, A., Grimshaw, M. and Williams, A. (2010) "Uncanny behavior in survival horror games," *Journal of Gaming and Virtual Worlds*, vol. 2, no. 1, pp. 3–25.

Tinwell, A., Grimshaw, M., Williams, A. and Abdel Nabi, D. (2011) "Facial expression of emotion and perception of the Uncanny Valley in virtual characters," *Computers in Human Behavior*, vol. 27, no. 2, pp. 741–749.

Topol, E. J. (2012) *The Creative Destruction of Medicine: How the Digital Revolution Will Create Better Health Care*, New York: Basic Books.

Tronick, E., Adamson, L., Als, H. and Brazelton, T. B. (1975) "Infant emotions in normal and perturbated interactions," Presentation at the biennial meeting of the Society for Research in Child Development, Denver, CO, April 1975.

Wells, S. (2011) *Mars Needs Moms* [Motion picture]. Burbank, CA: Walt Disney Pictures.

Young, J. (2011) "Did *Mars Needs Moms* sink Robert Zemeckis' 'Yellow Submarine'?" *Entertainment Weekly*. Retrieved May 8, 2014, from http://insidemovies.ew.com/2011/03/16/yellow-submarine-mars-needs-moms/.

Zemeckis, R. (2004) *The Polar Express* [Motion picture]. Los Angeles, CA: Castle Rock Entertainment.

Index

Printed and bound by CPI Group (UK) Ltd, Croydon, CR0 4YY

23/10/2024

01777708-0014